MyMathLab®
Notebook

Developmental Mathematics:
Basic Mathematics, Introductory Algebra, and Intermediate Algebra
Second Edition

John Squires
Chattanooga State Community College

Karen Wyrick
Cleveland State Community College

PEARSON

Boston Columbus Indianapolis New York San Francisco Upper Saddle River
Amsterdam Cape Town Dubai London Madrid Milan Munich Paris Montreal Toronto
Delhi Mexico City São Paulo Sydney Hong Kong Seoul Singapore Taipei Tokyo

ISBN-13: 978-0-321-98451-7
ISBN-10: 0-321-98451-X

1 2 3 4 5 6 CRK 18 17 16 15 14

www.pearsonhighered.com

PEARSON

Table of Contents

Name: _____   Date: _____

Instructor: _____   Section: _____

Notebook 1.1
Whole Numbers

_____ are the numbers
0, 1, 2, 3, 4, 5, 6, 7, 8, 9, 10, 11, 12, 13, 14, …

Note: How many digits are there? _____
        List them:

What is the difference between digits and whole numbers?

What system do we use to write whole numbers? _____ _____.

What does that mean? The _____ of each digit in a whole number depends
on the _____ as well as its _____ within the number.

Complete the following place value chart:

Each set of three place values forms a _____.

What is the difference in a digit's place value and its actual value?

EXAMPLE 1 Consider the number 13, 579.
 What digit is in the tens place value?

What is the place value of the 3?

What is the actual value of the 3?

We can use _____ to better understand the value of
each digit in a number.

Expanded notation shows the _____ _____ of each digit
based on its _____ _____.

1

EXAMPLES 2 & 3 Write the following numbers in expanded notation.

4295

245,103

What are the steps for converting numbers to words?

     Start at the _____.

     Write the number in each period , followed by _____

     Do this for each period, …

     Do not include the word _____

Note: True/False You must write the comma when writing a 4-digit number.

EXAMPLES 4 & 5 Convert the following numbers to words.

1209

821,504

A number written using digits rather than words is said to be

in _____ _____.

Write the steps for writing numbers in standard form.

     Read the number from _____ to _____.

     Write the number in each period, …

EXAMPLES 6 & 7 Write the following numbers in standard form.

twenty-four thousand, seven hundred twelve

Nine thousand, three hundred fourteen

CONCEPT CHECK: How many place values are in a period?

GUIDED EXAMPLES:

1. For 2,053,479, what is the place value of the 5?

2. Write in expanded notation. 14,295

3. Convert to words. 35,697

4. Convert to standard form. Three thousand, two hundred seventy-eight

2

Name: _____   Date: _____

Instructor: _____   Section: _____

Notebook 1.2
Rounding

What do we mean by rounding?

When we want to get an answer that is _____, but not

_____.

Rounding is finding a number _____

_____ but _____ to work with.

To round a number, what do you need to know?

EXAMPLE 1 Round 412 to the nearest ten.

EXAMPLE 2 Round 617,725 to the nearest thousand

What is the procedure for rounding?

1. _____ the digit in the place to which you are rounding.

2. Look at the digit to the _____ of the _____ digit.

3. If that digit is:

4 or less, _____ the underlined digit.
This is called _____ _____.

5 or more, _____ the underlined digit by _____.
This is called _____ _____.

4. Replace all digits to the right of the digit _____.

EXAMPLE 3 Round to the nearest thousand. 506,243.

EXAMPLE 4 Round to the nearest hundred. 101,697.

Note: Sometimes the digit to the left of the underlined digit must also be changed. When might this occur?

3

EXAMPLE 5 Round to the nearest ten. 1296

EXAMPLE 6 Round to the nearest hundred. 449,985

CONCEPT CHECK: When rounding a number to the nearest place value, and the number to the right of the digit being rounded is 5, do you round up or round down?

GUIDED EXAMPLES:
1. Round 457 to the nearest ten.

2. Round 19,256 to the nearest thousand.

3. Round 34,996 to the nearest ten.

4. Round 732,819 to the nearest ten thousand.

4

Name: _____   Date: _____

Instructor: _____   Section: _____

Notebook 1.3
Adding Whole Numbers; Estimation

_____ occurs when you combine numbers?

What is the answer to an addition problem?

EXAMPLES 1 & 2 Find the sum.

$5 + 6$

$7 + 2$

What does the Commutative Property of Addition state?
Then give an example.

What does the Associative Property of Addition state?
Then give an example.

What does the Addition Property of Zero state?
Then give an example.

EXAMPLES 3 – 6 For each statement, which property tells us the statement is true? The properties are Commutative and Associative Properties, and the Addition Property of Zero.

$2 + 4 = 4 + 2$ Property: _____

$(3 + 5) + 7 = 3 + (5 + 7)$ Property: _____

$6 + 0 = 6$ Property: _____

$(1 + 7) + 8 = 8 + (1 + 7)$ Property: _____

Explain the process for adding whole numbers with several digits.
*

*

*

5

EXAMPLES 7 & 8 Find the sum.

126 + 12

349 + 268 + 1345

What does it mean to estimate?

EXAMPLE 9 Estimate by rounding to the nearest hundred. Then compare the result with the exact answer.   7026 + 13,479

Estimate: Actual:

Write down some key words for phrases for addition:

EXAMPLES 10 - 12 Write the following using symbols, then simplify.

Find 6 increased by 5.

Find 7 more than 10.

Find the sum of 2 and 8.

CONCEPT CHECK: Can you name three words or phrases that indicate addition?

GUIDED EXAMPLES:

1. Add.   137 + 29

2. Add.   4853 + 6484

3. Estimate the total amount spent by rounding to nearest ten dollars.
    $526 + $841

4. Write in symbol form & simplify.
    four increased by six

Notebook 1.4
Subtracting Whole Numbers

_____ of two numbers is taking away one number or quantity from another.

What is the answer to a subtraction problem?

EXAMPLE 1 Consider the problem 6 – 2. What is the difference?

Write down the two subtraction facts. Give an example.
1. Any number minus itself is _____.

2. Any number minus zero is _____.

Do you know your addition facts? If not, now is the time to review.

True or False: Every subtraction problem has two related addition
                        problems.

EXAMPLE 2 Rewrite 12 – 5 = 7 as two addition problems.

EXAMPLES 3 & 4 Subtract each of the following. Check by adding.
         10 – 4

         17 – 2

Write down the steps for subtracting whole numbers with several digits.
         *

         *

         *

         *

         *

Borrowing is sometimes referred to as _____.

EXAMPLE 5 Subtract 57 – 38. Check by adding.

EXAMPLES 6 & 7 Subtract. Check by adding.
  232 – 141

  2010 – 1963

What is meant by estimation?

EXAMPLE 8 Estimate by rounding to the nearest thousand.
 Compare the estimate with the exact answer.   11,976 – 1245

   Estimate: Actual:

Write down some key words and phrases for subtraction:

CAUTION!! When translating words, be careful to write the numbers
 in the correct order. When should the order be written in opposite order?

EXAMPLES 9 & 10 Write the following using symbols, then simplify.
   Find 6 decreased by 5.

   Find 15 less than 51.

CONCEPT CHECK: Can you name three words or phrases that indicate
          subtraction?

GUIDED EXAMPLES:

1. Find the difference. 78 – 47

2. Subtract. 137 – 29

3. Estimate by rounding to nearest thousand.
   211,836 – 11,784

4. Write in symbols then simplify.
   6 subtracted from 12.

Name: _____   Date: _____

Instructor: _____   Section: _____

Notebook 1.5
Basic Problem Solving

What are the four steps for Problem Solving?

1.

2.

3.

4.

Write down some key words for addition:

Write down some key words for subtraction:

EXAMPLES 1 & 2 Translate the following to symbols and simplify.

The sum of 5 and 3

The difference of 6 and 2

The _____ of a figure is the distance around the figure.
How do you find the perimeter of a figure?

EXAMPLE 3 A rectangular garden measures 9 feet in length and is 5 feet wide. How many feet of fencing are needed to enclose the garden?

Understand the problem

Create a plan

Find the answer

Check the answer

EXAMPLE 4 A local community college requires 64 credit hours for an Associate Degree. Kylie earned 28 credit hours this past year. How many more credit hours does Kylie need in order to get her degree?

Understand the problem

Create a plan

Find the answer

Check the answer

EXAMPLE 5 On Monday, Alex opened a checking account with an initial deposit of $300. She bought groceries for $75, spent $25 on gas, and spend $20 on new clothes. How much money is in her account after these purchases?

Understand the problem

Create a plan

Find the answer

Check the answer

CONCEPT CHECK: What is the first step in problem solving?

GUIDED EXAMPLES:

1. How much money did Kristen make if she earned $100 but spent $20 and $10?

2. How much money does Jake receive in two weeks if he gets a weekly allowance of $10 and an additional $5 for mowing the lawn?

3. How much money do the Reyher's have left to spend on souvenirs if they started with $200 and bought tickets for $128, parking for $12 & food for $38?

Name: _____   Date: _____

Instructor: _____   Section: _____

Notebook 1.6
Multiplying Whole Numbers

_____ is a shortcut for repeated addition.
What are some ways in which we can write multiplication?

EXAMPLE 1 Tina is scheduled to work 3 days, 10 hours each day.

EXAMPLE 2 Pierre buys 4 dozen eggs at the store.

Do you know your basic multiplication facts?
If not, now is the time to review. Please look in the study guide for an empty multiplication table.

What is the answer to a multiplication problem?

The numbers being multiplied are called _____.

EXAMPLE 3 Identify the factors and find the product? 3×4
Factors: Product:

What does each of the following properties state? Give an example.
Commutative Property of Multiplication -

Associative Property of Multiplication -

Multiplication Property of Zero -

Multiplication Property of One -

Distributive Property of Multiplication over Addition (Distributive Property) –

EXAMPLES 4 & 5 Rewrite using the Distributive Property and simplify.
      $8(3 + 6)$ $2(9 + 2)$

Note: The Distributive Property can be used to make mental calculations.
EXAMPLE 6 Multiply $5(23)$.

11

EXAMPLES 7–11 For each statement, which property tells us the statement is true? Choose from the following:
Associative Property, Commutative Property, Distributive Property, Multiplication Property of Zero, and Multiplication Property of One.

$6 \cdot 25 = 25 \cdot 6$ Property: _____

$5(30 + 2) = 5(30) + 5(2)$ Property: _____

$0 \cdot 7 = 0$ Property: _____

$6 \cdot (4 \cdot 7) = (6 \cdot 4) \cdot 7$ Property: _____

$9 \cdot 1 = 9$ Property: _____

Write down the steps for multiplying whole numbers by a single-digit number.
1.

2.

3.

Note: Carrying may also be referred to as _____.

EXAMPLE 12 Calculate the following product. $3 \cdot 248$

Write down some key words and phrases for multiplication:

EXAMPLES 13 & 14 Write the following using symbols, then simplify.
     Find the product of 2 and 17.

     Find 721 times 9.

CONCEPT CHECK: Name two words or phrases that indicate multiplication?

GUIDED EXAMPLES:
1. Rewrite using the Distributive Property. Simplify: $3(8 + 2)$

2. Calculate the product: $157 \cdot 2$

3. Calculate the product: $3(243)$

4. Write in symbol form then simplify: the product of 4 and 5

Name: _____   Date: _____

Instructor: _____   Section: _____

Notebook 1.7
Dividing Whole Numbers

_____ is the operation of splitting a quantity or number into equal parts.

EXAMPLE 1 A group of 15 people goes on a whitewater rafting trip. There are 3 boats, so the group of 15 people needs to be divided into 3 groups of equal size. In this case, each boat will have 5 people, meaning that 15 divided by 3 is 5.

Do you know your basic multiplication facts? You will need them in order to divide. If not, now is the time to review.

Four Basic Division Facts: (Give an example of each.)
1. Any non-zero number divided by itself is _____.

2. Any number divided by 1 is _____.

3. Zero divided by any non-zero number is _____.

4. Division by 0 is _____.

What is the answer to a division problem?

The number being divided is the _____.

The number you are dividing by is called the _____.

What are 3 different ways a division problem can be written?

Write the 4 different ways division can be written in symbols.

EXAMPLE 2 Divide the following. $36 \div 4$

The number that is left over from a division problem is the _____.

A number divides exactly into another number if _____.

13

Write the steps for performing long division:

1. Set up the problem using the _____ _____ _____.

2. _____ the _____ into part of the _____.

3. _____ this result by the _____.

4.

5. _____ _____ the next digit from the _____.

6. Repeat until all digits in the _____ have been used.

EXAMPLES 3 & 4 Divide the following.

$$92 \div 4 \qquad\qquad 3\overline{)148}$$

Write down some key words and phrases for division:

CAUTION!! When translating for division, be careful to write down the numbers in the correct order.

EXAMPLES 5 & 6 Write the following using symbols, then simplify.

Find the quotient of 21 and 7.

Find 72 divided by 6.

CONCEPT CHECK: In the problem $\dfrac{6}{2} = 3$, which number is the quotient?

GUIDED EXAMPLES:

1. Divide.

$$56 \div 7$$

2. Divide.

$$\frac{0}{7}$$

3. Divide. Find the quotient and remainder.

$$68 \div 9$$

4. Write in symbol form, then simplify: the quotient of 24 and 3

14

Name: _____   Date: _____

Instructor: _____   Section: _____

Notebook 1.8
More with Multiplying and Dividing

Write down the steps for multiplying whole numbers with several digits.

1. _____ _____ _____ vertically so that the digits in the same place values line up.

2. Starting at the ones column, _____ the bottom digit by each digit in the _____ _____. Moving left, …

3. For each multiplication, start the _____ in the _____ _____ of the bottom digit. Place _____ to the _____ of each product to fill in all place values of that product.

4. If any product is _____ than _____, write down the _____ digit of that product. Carry the _____ digit to the next place value _____ .

5. Add the _____ _____ to get the overall _____.

EXAMPLES 1 & 2 Calculate the following products.

$$21 \cdot 17 \qquad\qquad 32 \cdot 125$$

What are the steps for multiplying numbers ending in zero(s)?

 1.

2.

3.

EXAMPLE 3 Multiply.   260(400)

EXAMPLE 4   Estimate by rounding to the nearest ten.   7227(87)

15

What are the steps for performing long division?
1.

2.

3.

4.

5.

6.

Note: The final remainder should be less than the _____.

EXAMPLE 5 Divide the following. $1482 \div 12$

EXAMPLE 6 Divide the following. $4725 \div 225$

CONCEPT CHECK:
How many zeros are in the final product of 55,000 and 300?

GUIDED EXAMPLES:
Calculate the following products.
1.   157(62)

2. 5700 · 300

3. Estimate by rounding to the nearest ten.   634 · 27

4. Divide. $304 \div 23$

16

Name: _____   Date: _____

Instructor: _____   Section: _____

Notebook 1.9
Exponents

An _____ is used as a shortcut for repeated multiplication.

What does 2^3 mean?

Which is the base? _____   Which is the exponent? _____

CAUTION!!

What is the number, or factor, being multiplied by itself?

The _____ is the number of times the base is used as a factor.

EXAMPLE 1 Identify the base and exponent, then evaluate.
5^2 Base: _____   Exponent: _____   Value:

Note: For an exponent of 2, we say "_____".

        For an exponent of 3, we say "_____".

EXAMPLES 2 & 3 Evaluate (find the values of) the following.
     1^4

     6^2

EXAMPLES 4 & 5 Write the following using exponents,
then evaluate.
     $(3)(3)(3)$

     $7 \cdot 7$

Rule: Any number raised to the power of 1 is _____.

EXAMPLES 6 & 7 Evaluate.
     7^1 13^1

17

Rule: Any non-zero number raised to the 0 power is _____.

EXAMPLES 8 & 9 Evaluate.

5^0 79^0

EXAMPLES 10 & 11 Evaluate.

8^2 31^0

Note: Small exponents can result in _____ very quickly.

EXAMPLE 12 Calculate 10^2, 10^3 and 10^6.

10^2 10^3 10^6

This rapid growth, called _____, occurs in science and business.

CONCEPT CHECK: How can 2^5 be written without exponents?

GUIDED EXAMPLES:

1. Identify the base and exponent.
   5^4

2. Evaluate.
   2^5

3. Write using exponents and evaluate.
   $3 \cdot 3 \cdot 3 \cdot 3$

4. Write using exponents and evaluate.
   $(12)(12)$

Name: _____   Date: _____

Instructor: _____   Section: _____

Order of Operations and Whole Numbers

OTHER NOTES

When there is more than one operation involved, use the Order of
Operations so that you will know the order in which to perform the operations.

Write down the steps for using the order of operations:

P:

E:

MD:

AS:

What can you help to remember this order?

EXAMPLE 1 Calculate.

$4 + 6 \cdot 3$

EXAMPLES 2 & 3 Simplify by using the order of operations.

$7 + (13 - 9)$ $\qquad\qquad$ $3^2 + 5 \cdot 4$

P: $\qquad\qquad\qquad\qquad$ P:

E: $\qquad\qquad\qquad\qquad$ E:

MD: $\qquad\qquad\qquad\qquad$ MD:

AS: $\qquad\qquad\qquad\qquad$ AS:

EXAMPLES 4 & 5 Simplify.

$3(4 - 2)^2 + 5$ $\qquad\qquad$ $24 \div 3 \cdot 2$

P:

E:

MD:

AS:

Note: What are some other symbols that act like parentheses?

When you see a fraction bar, act like there are parentheses around the _____ and _____.

For example, what does $\dfrac{2+1}{4+3}$ mean?

EXAMPLE 6 Simplify. $\dfrac{22+10}{4\cdot 5-4}$

CONCEPT CHECK:

What operation should be performed first in simplifying $9-5+16\div 2^2$?

GUIDED EXAMPLES:

Simplify.

1. $2+8\times 3-1$

2. $12\div (5-2)+7$

3. $5+(8-2)^2+4$

4. $\dfrac{20+20}{6+4}$

Name: _____   Date: _____

Instructor: _____   Section: _____

Notebook 1.11
More Problem Solving

OTHER NOTES

Recall the four steps for Problem Solving.
1.

2.

3.

4.

Recall the key words and phrases for:
Addition:

Subtraction:

Multiplication:

Division:

EXAMPLES 1–3 Translate the following to symbols.
Then simplify using the order of operations.

twice the sum of 7 and 4

the difference of 9 and 3, divided by 3

three times the product of 2 and 4

The _____ of a _____ is the measure of the surface inside a figure.

How do you find the area of a rectangle?

EXAMPLE 4 Find the area of the rectangular room that has a length of 21 feet and a width of 14 feet.

Understand the problem

Create a plan

Find the answer

Check the answer

EXAMPLE 5 Desmond makes a salary of $39,000 per year. What is his monthly salary?

Understand the problem

Create a plan

Find the answer

Check the answer

EXAMPLE 6 The phone company charges $20 for installation and $25 a month for basic service. How much will Monique pay for one year of service with installation?

Understand the problem

Create a plan

Find the answer

Check the answer

CONCEPT CHECK:
What two steps need to be done before calculation in solving a problem?

GUIDED EXAMPLES:
Translate to symbols, then simplify.
1. 2 more than the product of 3 and 5

2. twice the product of 3 and 7

3. Michelle makes a salary of $54,000 per year. What is her monthly salary?

Name: _____   Date: _____

Instructor: _____   Section: _____

Notebook 2.1
Factors

A _____ is a number that is multiplied by another number.

Also, a factor is a whole number that _____ _____ into another number.

What are the factors of 6?

EXAMPLES 1 & 2 List all factors of 8.

List all factors of 24. Hint: Use factor pairs.

A _____ _____ is a number that divides exactly into two or more numbers. Can there be more than one?

What is the largest number that divides exactly into two or more numbers?

What are the steps finding the greatest common factor?
1. List all _____.
2. Identify the factors that are _____.
3. Choose the _____ of these _____ _____.

EXAMPLE 3 Find the greatest common factor, or GCF of 15 and 24.
Factors of 15:

Factors of 24:

Common factors:

Largest:

EXAMPLE 4 Find the greatest common factor, or GCF of 18 and 54.
Factors of 18:

Factors of 54:

Common factors:

Largest:

Rules for Divisibility:

A number is divisible by:	If:	Example:
2	It ends in ...	
3	The sum of ...	
5	It ends in ...	
6	It is divisible by both ...	
9	The sum of ...	
10	It ends in ...	

EXAMPLE 5 Find the GCF of 132 and 198.

Factors of 132:

Factors of 198:

Common factors:

Largest:

What does "to factor" mean?

EXAMPLE 6 List three ways to factor 18.

CONCEPT CHECK: What is a factor?

GUIDED EXAMPLES:

1. Is 1695 divisible by 2? 3? 5? 9? 10?

2. Find the GCF of 12 and 16.

3. Find the GCF of 24 and 56.

4. List as least 2 ways to factor 36.

Name: _____   Date: _____

Instructor: _____   Section: _____

Notebook 2.2
Prime Factorization

Factor (noun) – a number that is being _____.
Factor (verb) – to write as a _____.

EXAMPLE 1 a. Find all factors of 36.   b. Factor 36 several ways.

A _____ _____ is a whole number greater than 1 that has exactly two factors, the number itself and 1.

Is 0 prime?   Is 1 prime?

Prime Numbers:

Whole Number	Factors	Prime??
2		
3		
4		
5		
6		
7		
8		
9		

Write down the first 15 prime numbers:

_____ _____ are whole numbers greater than 1 that are not prime.

Is 0 composite?   Is 1 composite?

True/False: Every number is either prime or composite.

EXAMPLE 2 Identify each whole number as prime, composite, or neither.
        2, 8, 1, 13, 22, 55, 29, 37, 99

The _____ _____ of any whole number is the factored form in which all factors are prime numbers.

OTHER NOTES

What are the steps for finding the prime factorization?

1. Write the number as a _____ of _____.

2. Check to see if each of these factors is _____.
   If so, _____; if not, write it as a _____.

3. Continue to do this until all factors are _____.

4. Write the prime factorization as a product of all the _____
   _____ _____.

EXAMPLE 3 Find the prime factorization of 24.

EXAMPLE 4 Find the prime factorization of 60.

Note: If factors are listed from smallest to largest, every composite number's prime factorization can be written in _____.

CONCEPT CHECK: How many factors does a prime number have?

GUIDED EXAMPLES:

1. Identify as prime, composite, or neither. 55 _____ 17 _____

2. Find the prime factorization of 35.

3. Find the prime factorization of 48.

4. Find the prime factorization of 28.

Name: _____   Date: _____

Instructor: _____   Section: _____

Notebook 2.3
Understanding Fractions

What is the difference between whole numbers and fractions?
Give examples of each.

What are numbers that represent the parts of a whole?

EXAMPLE 1 A pizza has 8 slices. Sam ate 3 of the 8 slices for lunch.

The top number of a fraction is called the _____.

The bottom number of a fraction is called the _____.

Which one indicates the number of equal parts being considered?

Which one indicates the number of parts in the whole?

EXAMPLE 2 Write a fraction that best represents the shaded part.

a.   b.

c.

What kind of fraction has the numerator less than the denominator?
Proper or improper??
A proper fraction has a value _____ _____ 1.

Write some examples of proper fractions:

What kind of fraction has the numerator greater than or equal to
the denominator? Proper or improper??

An improper fraction has a value _____.

Write some examples of improper fractions:

27

Note: Any whole number greater than or equal to 1 can be written as an
_____ _____ with the number in the numerator and a
_____ in the denominator.

EXAMPLES 3–7 Identify each fraction as proper or improper.

$$\frac{4}{9}$$

$$\frac{11}{4}$$

$$\frac{7}{6}$$

$$\frac{3}{3}$$

$$\frac{0}{5}$$

Four Basic Facts about Fractions:
1. Any non-zero number divided by itself is _____.
    Ex.

2. Any number divided by 1 is _____.
    Ex.

3. Zero divided by any non-zero number is _____.
    Ex.

4. Division by 0 is _____.
    Ex.

CONCEPT CHECK: Division by zero is _____.

GUIDED EXAMPLES:
1. Write a fraction that represents the shaded part.

2. Identify $\frac{0}{7}$ as proper or improper

3. Identify $\frac{4}{4}$ as proper or improper.

4. Simplify $\frac{11}{0}$.

Name: _____   Date: _____

Instructor: _____   Section: _____

Notebook 2.4
Simplifying Fractions—GCF and Factors Method

Every fraction can be written in many ways.
Write several different ways to represent one-half.

What are equivalent fractions?

A fraction is said to be in _____ _____
if there is _____ _____ _____ other than 1
that divides exactly into the _____ and _____.

What are the steps to simplify a fraction using the GCF method?
1. Find the _____ of the numerator and denominator.

2. Divide the _____ and _____ by the _____.

True/False: Dividing the numerator and the denominator by
the same non-zero number does not change the value of a fraction.

EXAMPLE 1 Write in simplest form using the GCF method. $\dfrac{9}{15}$

Find the GCF:

Divide num and den by GCF:

EXAMPLE 2 Write in simplest form using the GCF method. $\dfrac{72}{96}$

Find the GCF:

Divide the num and den by GCF:

If you are having a hard time finding the GCF, use any common factor.

What are the steps for simplifying fractions using the factors method?
1. Find any factor that is _____.

2. Divide the numerator and denominator by _____.

3. Repeat this until there are _____.

29

EXAMPLE 3 Write in simplest form using the factors method. $\dfrac{20}{32}$

EXAMPLE 4 Write in simplest form using the factors method. $\dfrac{72}{96}$

CONCEPT CHECK: To simplify $\dfrac{10}{20}$ using the GCF method, what number will be divided into the numerator and denominator?

GUIDED EXAMPLES:

Write in simplest form using the GCF method.

1. $\dfrac{12}{16}$

2. $\dfrac{75}{125}$

Write in simplest form using the factors method.

3. $\dfrac{30}{75}$

4. $\dfrac{24}{54}$

Name: _____   Date: _____

Instructor: _____   Section: _____

Notebook 2.5
Simplifying Fractions—Prime Factors Method

We can use prime factorizations to write a fraction in simplest form.

What are the steps for simplifying fractions using the
Prime Factors Method?
1. Write the numerator as a _____.
2. Write the denominator as a _____.
3. Divide by _____ in the numerator
   and denominator.
4. Multiply the _____ _____ to get simplest form.

Note: What results when you divide the numerator and denominator by the
same non-zero number?

EXAMPLE 1 Write in simplest form using the prime factors method. $\dfrac{24}{36}$

Write numerator as a product of primes:

Write denominator as a product of primes:

Divide numerator and denominator
by common factors:

Multiply the remaining factors:

EXAMPLE 2 Write in simplest form using the prime factors method. $\dfrac{84}{91}$

Write numerator as a product of primes:

Write denominator as a product of primes:

Divide numerator and denominator
by common factors:

Multiply the remaining factors:

OTHER NOTES

EXAMPLE 3 Write in simplest form using the prime factors method. $\dfrac{39}{195}$

Write numerator as a product of primes:

Write denominator as a product of primes:

Divide numerator and denominator by common factors:

Multiply the remaining factors:

CONCEPT CHECK: When dividing the numerator and denominator of a fraction by the same non-zero number, what happens to the value of the original fraction?

GUIDED EXAMPLES:

1. Write in simplest form using the prime factors method. $\dfrac{25}{35}$

2. Write in simplest form using the prime factors method. $\dfrac{84}{90}$

3. Write in simplest form using the prime factors method. $\dfrac{42}{126}$

Notebook 2.6
Multiplying Fractions

How do you find $\dfrac{1}{2}$ of $\dfrac{2}{3}$?

Show this.

What are the steps for multiplying fractions?
1.
2.
3.

EXAMPLE 1 Multiply. Write your answer in simplest form. $\dfrac{4}{7} \cdot \dfrac{3}{5}$

Multiply numerators.

Multiply denominators.

Simplify.

EXAMPLE 2 Multiply. Write your answer in simplest form. $\dfrac{7}{10} \cdot \dfrac{5}{8}$

Multiply numerators.

Multiply denominators.

Simplify.

What are the steps for multiplying fractions by simplifying first?
1.
2.
3.

EXAMPLE 3 Multiply by simplifying first. $\dfrac{7}{10} \cdot \dfrac{5}{8}$

Write as one fraction.

Divide by common factors.

Multiply the remaining factors.

EXAMPLE 4 Multiply by simplifying first. $\dfrac{4}{15} \cdot \dfrac{9}{16}$

EXAMPLE 5 Multiply by simplifying first. $3 \cdot \dfrac{2}{9}$

EXAMPLE 6 Elizabeth bought a sandwich and ate half of it. She gave her brother one-half of what was left. How much of the sandwich did her brother get?

Understand the problem.

Create a plan.

Find the answer.

Check the answer.

EXAMPLE 7 Multiply by simplifying first. $\dfrac{2}{8} \cdot \dfrac{5}{15}$

CONCEPT CHECK: You multiplied fractions to get $\dfrac{33}{48}$. What will your final answer be?

GUIDED EXAMPLES:

1. Multiply $\dfrac{2}{3} \cdot \dfrac{4}{9}$. Write your answer in simplest form.

2. Multiply $\dfrac{3}{10} \cdot \dfrac{5}{24}$ by simplifying first.

3. Multiply $\dfrac{2}{7} \cdot \dfrac{21}{50}$ by simplifying first.

4. Multiply $\dfrac{3}{8} \cdot 2$ by simplifying first.

Notebook 2.7
Dividing Fractions

How do you know if two numbers are reciprocals?

How do you find the reciprocal of a fraction?

How do you find the reciprocal of a whole number?

EXAMPLES 1 & 2 Find the reciprocals of the following numbers.

$\frac{3}{4}$

5

What is the reciprocal of 0?

What happens when you divide a number by 2 and multiply the same number by $\frac{1}{2}$?

Rule: Dividing by a fraction is the same as _____.

What are the steps for dividing fractions?
1.
2.
3.
4.

EXAMPLE 3 Divide. Write your answer in simplest form. $\frac{3}{5} \div \frac{1}{2}$

Invert the second fraction.

Multiply by the first fraction.

Divide by common factors.

Multiply the remaining factors.

EXAMPLES 4 & 5 Divide. Write your answers in simplest form.

$$\frac{4}{5} \div \frac{3}{8}$$

$$\frac{6}{7} \div \frac{3}{2}$$

.

EXAMPLES 6 & 7 Divide. Write your answers in simplest form.

$$12 \div \frac{2}{3}$$

$$\frac{6}{7} \div 9$$

CONCEPT CHECK: How would you rewrite $\frac{3}{8} \div \frac{4}{9}$ as a multiplication

problem?

GUIDED EXAMPLES:
Divide each of the following. Write your answer in simplest form.

1. $\dfrac{2}{3} \div \dfrac{4}{5}$

2. $\dfrac{4}{7} \div \dfrac{8}{11}$

3. $\dfrac{9}{10} \div 6$

4. $\dfrac{5}{12} \div \dfrac{15}{16}$

Name: _____ Date: _____

Instructor: _____ Section: _____

Notebook 2.8
Applications of Multiplying and Dividing Fractions

Write the steps for multiplying fractions by simplifying first.

1.
2.
3.

EXAMPLE 1 Multiply $\dfrac{3}{5} \cdot \dfrac{7}{9}$. Write your answer in simplest form.

What are the steps for dividing fractions?

1.
2.
3.
4.

EXAMPLE 2 Divide $\dfrac{2}{7} \div \dfrac{4}{5}$. Write your answer in simplest form.

EXAMPLE 3 Sam received $1200 through an academic scholarship. He used two-thirds of it for tuition. How much of the scholarship money was used for tuition?

Understand the problem

Create a plan

Find the answer

Check the answer

EXAMPLE 4 Allen has five-sixths of a pound of small pebbles that are used to cover the bottom of fish bowls. How many fish bowls can he prepare if each bowl uses one-twelfth of a pound of the pebbles?

Understand the problem

Create a plan

Find the answer

Check the answer

37

EXAMPLE 5 One-fourth of the members of the local chamber choir sing soprano. Of these sopranos, four-fifths of them are female. What fraction of the choir members are female and sing soprano?

Understand the problem

Create a plan

Find the answer

Check the answer

EXAMPLE 6 Megan has three-fourths of a gallon of tea and needs to fill as many glasses as possible. If each glass holds two-thirds of a cup, how many glasses can she fill? Hint: one gallon is the same as 16 cups.

Understand the problem

Create a plan

Find the answer

Check the answer

CONCEPT CHECK: What operation is required to solve this problem? A large recipe that makes 24 servings calls for $\frac{2}{3}$ cups of oil. Andrea plans to make $\frac{1}{4}$ of the recipe.

GUIDED EXAMPLES:

1. Ellie needs to earn 66 credit hours at Roane Mountain College in order to get an Associates degree. She will earn $\frac{1}{6}$ of them this semester. How many credits will she earn this semester?

2. A dump truck holds three-fourths of a ton of mulch. How many flowerbeds can be covered if each flowerbed needs one-eighth of a ton of mulch?

3. Two-thirds of Kylie's allowance is saved for a down payment on a new car. Her parents will match one-fourth of this saved amount. What fraction of Kylie's allowance will her parents match?

4. A cookie recipe calls for three-fourths of a cup of peanut butter. How much peanut butter is needed to make one-half of the recipe?

Notebook 3.1
Finding the LCM—List Method

EXAMPLE 1 Find the first five multiples of 9.

EXAMPLE 2 Find the first four multiples of 20.

A _____ _____ of two or more numbers is a number that is a multiple of the given numbers.

True/False You can find a common multiply by multiplying the given numbers together.

The _____ _____ _____ of two or more numbers is the _____ number that is a multiple of the given numbers.

What are the steps for finding the LCM of two or more numbers using the List Method?
1.
2.
3.

EXAMPLE 3 Find the least common multiple of 9 and 12.
List the first several multiples of 9.

List the first several multiples of 12.

Identify common multiples.

Choose the least.

EXAMPLE 4 Find the least common multiple of 6, 8, and 12.

List the first several multiples of 6.

List the first several multiples of 8.

List the first several multiples of 12.

Identify common multiples.

Choose the least.

CONCEPT CHECK: What are multiples of 8?

GUIDED EXAMPLES:
1. List the first five multiples of 16.

2. Find the least common multiple of 4 and 6.

3. Find the least common multiple of 12 and 15.

4. Find the least common multiple of 6, 9, and 18.

Notebook 3.2
Finding the LCM—GCF Method

What are the steps for finding the LCM of two numbers using the
GCF Method?
1.
2.
3.

EXAMPLE 1 Find the LCM of 6 and 9 using the GCF Method.
Find the GCF of 6 and 9.
Multiply 6 and 9.
Divide this product by the GCF.

EXAMPLE 2 Find the LCM of 6 and 15 using the GCF Method.
Find the GCF of 6 and 15.
Multiply 6 and 15.
Divide this product by the GCF.

EXAMPLE 3 Find the LCM of 4 and 18 using the GCF Method.

EXAMPLE 4 Find the LCM of 7 and 9 using the GCF Method.

41

Note: If the GCF is 1, then the LCM is the
_____.

CONCEPT CHECK: Explain how to find the LCM of 10 and 25.

GUIDED EXAMPLES:

1. Find the LCM of 6 and 10 using the GCF method.

2. Find the LCM of 12 and 16 using the GCF method.

3. Find the LCM of 8 and 15 using the GCF method.

Name: _____   Date: _____

Instructor: _____   Section: _____

Notebook 3.3
Finding the LCM—Prime Factor Method

What is prime factorization of any whole number?

EXAMPLES 1 & 2

Find the prime factorization of 12.

Find the prime factorization of 42.

What are the steps for finding the LCM of two numbers using the prime factor method?

1.
2.
3.
4.

EXAMPLE 3 Find the LCM of 12 and 42 using the Prime Factor Method.
Prime factorization of 12:

Prime factorization of 42:

LCM:

EXAMPLE 4 Find the LCM of 8 and 30.
Prime factorization of 8:

Prime factorization of 30:

Write the prime factorizations
putting the common factors below
each other:

Write down the prime factor from
each column:

Multiply the lists of primes to get LCM:

43

EXAMPLE 5 Find the LCM of 18 and 24.

EXAMPLE 6 Find the LCM of 13 and 19.

CONCEPT CHECK: Write the prime factorization of 24.

GUIDED EXAMPLES:
Find the LCM of each pair of numbers using the prime factor method.
1. 15 and 21

2. 20 and 45

3. 72 and 84

Name: _____   Date: _____

Instructor: _____   Section: _____

Notebook 3.4
Writing Fractions with an LCD

OTHER NOTES

Things to consider when finding the LCM:
1. If one number divides exactly into the other, then the LCM is …

Give an example:

2. If the numbers have no common factor other than 1, then the LCM is …

Give an example:

3. If the two numbers have a common factor other than 1, then …

Give an example:

What are some methods that can be used to find the LCM?

What is the difference between LCM and LCD?

What is the least common denominator (LCD)?

EXAMPLES 1 & 2

Find the LCM of 4 and 12.

Find the LCD of $\dfrac{3}{4}$ and $\dfrac{5}{12}$.

True/False The same process is used to find both the LCD and the LCM.

Fractions are _____ _____ if they represent the same value.

Name some other fractions that represent $\dfrac{1}{6}$.

45

What are the steps for writing an equivalent fraction?

EXAMPLE 3 Write a fraction equivalent to $\dfrac{1}{6}$ using a denominator of 12.

List the steps for writing fractions with an LCD.

1.

2.

3.

EXAMPLE 4 Rewrite $\dfrac{1}{6}$ and $\dfrac{7}{9}$ using the LCD as the denominator.

EXAMPLE 5 Rewrite $\dfrac{4}{15}$ and $\dfrac{5}{6}$ using the LCD as the denominator.

CONCEPT CHECK:

After finding the LCD of $\dfrac{1}{6}$ and $\dfrac{2}{7}$, what number will we multiply 6 by to get an equivalent fraction?

GUIDED EXAMPLES:

1. Rewrite $\dfrac{5}{8}$ using 24 as the denominator.

2. Rewrite $\dfrac{11}{12}$ and $\dfrac{7}{18}$ using the LCD as the denominator.

3. Rewrite $\dfrac{2}{3}$ and $\dfrac{3}{7}$ using the LCD as the denominator.

Name: _____   Date: _____

Instructor: _____   Section: _____

Notebook 3.5
Adding and Subtracting Like Fractions

Fractions with the same, or common, denominator are called
_____ _____.

What are unlike fractions?

EXAMPLES 1 & 2

$\dfrac{1}{8}, \dfrac{3}{8},$ and $\dfrac{7}{8}$ are _____ fractions.

$\dfrac{3}{7}$ and $\dfrac{3}{5}$ are _____ fractions.

Write the steps for adding like fractions:
1.
2.
3.

EXAMPLES 3 & 4 Add the following like fractions. Simplify if possible. .

$\dfrac{2}{5} + \dfrac{1}{5}$

$\dfrac{3}{10} + \dfrac{3}{10}$

Write the steps for subtracting like fractions:

1.
2.
3.

EXAMPLES 5 & 6 Subtract the following like fractions. Simplify if possible.

$\dfrac{6}{7} - \dfrac{2}{7}$

$\dfrac{11}{14} - \dfrac{5}{14}$

EXAMPLES 7 & 8 Add or subtract the following like fractions. Simplify if possible.

$$\frac{4}{11}+\frac{3}{11}+\frac{5}{11}$$

$$\frac{7}{8}-\frac{3}{8}$$

EXAMPLE 9 Shauna ate $\frac{1}{8}$ of a pizza and Kenzi ate $\frac{3}{8}$ of the same pizza. Together how much of the pizza did the two girls eat?
Understand the problem.

Create a plan.

Find the answer.

Check the answer.

CONCEPT CHECK: Danette walked to the store. First she walked down a street that is $\frac{2}{10}$ of a mile long, and then down another street that is $\frac{5}{10}$ of a mile long. Create a plan to show how far she walked to the store.

GUIDED EXAMPLES:

1. Subtract the like fractions $\frac{7}{10}-\frac{4}{10}$. Simplify if possible.

2. Add the like fractions $\frac{2}{15}+\frac{8}{15}$. Simplify if possible.

3. Subtract the like fractions $\frac{7}{9}-\frac{4}{9}$. Simplify if possible.

4. Add the like fractions $\frac{7}{12}+\frac{7}{12}+\frac{1}{12}$. Simplify if possible.

Notebook 3.6
Adding and Subtracting Unlike Fractions

What is the first thing you must do to add or subtract unlike fractions?

Caution!! You must have a _____ _____
when _____ or _____ fractions.

Write down the steps for adding and subtracting unlike fractions:
1.

2.

3.

4.

5.

EXAMPLE 1 Add $\dfrac{3}{4}+\dfrac{1}{6}$. Simplify if possible.

Find the LCD.

Rewrite each fraction with
the LCD as the denominator.

Add the numerators.
Keep the denominator.
Simplify if possible.

EXAMPLE 2 Add $\dfrac{3}{8}+\dfrac{2}{5}$. Simplify if possible.

EXAMPLE 3 Subtract $\dfrac{5}{6}-\dfrac{1}{3}$. Simplify if possible.

EXAMPLE 4 Subtract $\dfrac{7}{10} - \dfrac{4}{15}$. Simplify if possible.

EXAMPLE 5 Paul ate $\dfrac{3}{8}$ of an apple pie. His brother Jeff ate $\dfrac{1}{4}$ of the same pie.

How much more pie did Paul eat than Jeff?

Understand the problem

Create a plan

Find the answer

Check the answer

CONCEPT CHECK:

For the problem $\dfrac{5}{9} - \dfrac{1}{3}$, what is the first step?

GUIDED EXAMPLES:

1. Add $\dfrac{3}{7} + \dfrac{2}{5}$. Simplify if possible.

2. Add $\dfrac{1}{14} + \dfrac{3}{7}$. Simplify if possible.

3. Subtract $\dfrac{4}{9} - \dfrac{1}{12}$. Simplify if possible.

4. Subtract $\dfrac{14}{15} - \dfrac{3}{5}$. Simplify if possible.

Name: _____   Date: _____

Instructor: _____   Section: _____

Notebook 3.7
Order of Operations and Fractions

Recall: When there is more than one operation involved, use the Order of Operations so that you will know the order in which to perform the operations.

Write down the steps for using the order of operations:
P:

E:

MD:

AS:

Remember, certain symbols can act like parentheses. What are they?

EXAMPLE 1 Simplify by using the order of operations. $\left(\dfrac{1}{3}-\dfrac{1}{6}\right)+\dfrac{1}{2}$

P:

E:

MD:

AS:

Simplify:

EXAMPLE 2 Simplify. $\left(\dfrac{2}{5}+\dfrac{1}{5}\right)^{2}\div\dfrac{3}{10}$

P:

E:

MD:

AS:

Simplify:

EXAMPLE 3 Simplify. $\dfrac{1}{4} \div \dfrac{3}{2} + \dfrac{1}{2} \cdot \dfrac{1}{3}$

P:

E:

MD:

AS:

Simplify:

EXAMPLE 4 Simplify by using the order of operations.

$$\left(\dfrac{3}{4}\right)\left(\dfrac{1}{2} - \dfrac{1}{4}\right)^2 + \dfrac{2}{5} \cdot \dfrac{1}{2}$$

P:

E:

MD:

AS:

Simplify:

CONCEPT CHECK:

Given the problem, $\dfrac{1}{2}\left(\dfrac{3}{4} - \dfrac{1}{8}\right) + \left(\dfrac{1}{5}\right)^2$ which operation is performed first?

GUIDED EXAMPLES:
Simplify each of the following.

1. $\dfrac{3}{5} + \dfrac{1}{4} \cdot \dfrac{2}{3} - \dfrac{1}{10}$

2. $\dfrac{2}{3} \div \left(\dfrac{5}{8} + \dfrac{1}{2}\right) - \dfrac{1}{9}$

3. $\dfrac{5}{6} \cdot \dfrac{1}{2} + \left(\dfrac{2}{3} \div \dfrac{4}{9}\right)^2$

Name: _____ Date: _____

Instructor: _____ Section: _____

Notebook 3.8
More Applications of Adding and Subtracting Fractions

OTHER NOTES

Write the steps for adding and subtracting unlike fractions.

1.

2.

3.

4.

5.

EXAMPLE 1 Add $\dfrac{3}{5} + \dfrac{2}{9}$. Write your answer in simplest form.

EXAMPLE 2 Gabriel spends $\dfrac{1}{3}$ of his monthly paycheck on rent and

$\dfrac{3}{8}$ of his paycheck on utilities (electric, water, internet, and cell phone).

What fraction of his paycheck is spend on both rent and utilities?

Understand the problem

Create a plan

Find the answer

Check the answer

EXAMPLE 3 Helena is selling flowers for homecoming. The florist providing the flowers is to receive one-fourth of the total sales. Helena will receive three-fifths of total sales as her pay, but she needs to pay the florist from her share. How much of her original pay will Helena get to keep?

Understand the problem

Create a plan

Find the answer

Check the answer

EXAMPLE 4 In a class of 18 students, $\frac{1}{3}$ of them earned an A, $\frac{1}{2}$ of them earned a B, and $\frac{1}{6}$ of them earned a C. What fraction of the students earned an A or a B?

Understand the problem

Create a plan

Find the answer

Check the answer

EXAMPLE 5 Bradley Parks is developing a wildflower park that covers three-eighths of an acre. Enough wildflower seed has been purchased to see one-fourth of an acre. How much more of the park still needs to be seeded?

Understand the problem

Create a plan

Find the answer

Check the answer

CONCEPT CHECK: What is the first step in solving the following application? Last weekend it rained $\frac{1}{5}$ of an inch on Saturday and $\frac{7}{9}$ of an inch on Sunday. What was the total rainfall for the weekend?

GUIDED EXAMPLES:

1. Susie is making dinner and needs $\frac{5}{8}$ of a cup of shredded cheese for the casserole and $\frac{1}{3}$ of a cup of shredded cheese for the salad. How much shredded cheese will she need in all?

2. Jeremiah has three-fourths of a gallon of milk and Maggie has two-thirds of a gallon of milk. How much more milk does Jeremiah have than Maggie?

3. On Monday, we received $\frac{3}{8}$ of an inch of rain. On Tuesday, it rained an additional $\frac{1}{4}$ of an inch. What was the total rainfall on Monday and Tuesday?

Name: _____   Date: _____

Instructor: _____   Section: _____

Notebook 4.1
Changing a Mixed Number to an Improper Fraction

What is a mixed number?

An improper fraction has a _____ that is
_____ _____ or _____ the
denominator. It represents at least _____.

Write the following as a mixed number and an improper fraction.

Write down the steps for changing mixed numbers to improper fractions.
1.

2.

3.

EXAMPLE 1 Write $2\frac{1}{3}$ as an improper fraction.

EXAMPLES 2 & 3 Write each as an improper fraction.

$4\frac{5}{7}$ $6\frac{9}{10}$

Write the following picture as a whole number and an improper fraction.

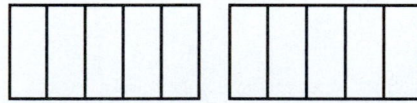
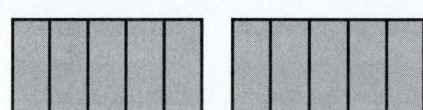

55

Write down the steps for changing whole numbers into improper fractions.

EXAMPLE 4 Write as an improper fraction. 25

EXAMPLE 5–7 Write each as an improper fraction.

$$20\frac{3}{10}$$

91

$$2\frac{37}{50}$$

CONCEPT CHECK: What is the process for converting 22 to an improper fraction?

GUIDED EXAMPLES:
Write as an improper fraction.

1. $5\frac{3}{8}$

2. $6\frac{2}{3}$

3. $18\frac{4}{5}$

4. 219

Name: _____   Date: _____

Instructor: _____   Section: _____

Notebook 4.2
Changing an Improper Fraction to a Mixed Number

Fractions can be represented by _____.

EXAMPLE 1 Write $\dfrac{13}{5}$ as a division problem.

Represent this visually:

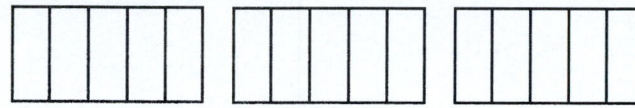

Write the steps for changing improper fractions to mixed numbers.

1.

2. Quotient –

   Remainder –

   Denominator -

3.

CAUTION!! Which part of the fraction goes on the outside of the division bar?

Which part of the fraction goes on the inside?

EXAMPLES 2 & 3 Write each as a mixed number.

$\dfrac{13}{4}$

$\dfrac{7}{5}$

Note: What happens if the remainder is 0 when changing improper fractions to mixed numbers?

EXAMPLES 4 & 5 Write each as a mixed or whole number.

$$\frac{19}{5}$$

$$\frac{28}{7}$$

CONCEPT CHECK:

Every fraction can be written as a division problem. How do you write $\frac{33}{4}$ as a division problem?

GUIDED EXAMPLES:

1. Write $\frac{35}{4}$ as a mixed number.

2. Write $\frac{60}{8}$ as a mixed number.

3. Write $\frac{56}{7}$ as a mixed number.

4. Write $\frac{99}{11}$ as a mixed number.

Name: _____   Date: _____

Instructor: _____   Section: _____

Notebook 4.3
Multiplying Mixed Numbers

Write the steps for multiplying mixed numbers:

1.

2.

3.

EXAMPLE 1 Multiply.

$$1\frac{1}{2} \cdot 5\frac{1}{6}$$

EXAMPLE 2 Multiply.

$$1\frac{2}{3} \cdot 4\frac{1}{5}$$

Write as improper fractions:

Multiply the fractions:

Rewrite as a mixed or whole number:

EXAMPLE 3 Multiply.

$$3\frac{1}{4} \cdot 2\frac{3}{5}$$

Write as improper fractions:

Multiply the fractions:

Rewrite as a mixed or whole number:

EXAMPLE 4 Multiply.

$$2\left(3\frac{1}{2}\right)\left(1\frac{2}{7}\right)$$

Write as improper fractions:

Multiply the fractions:

Rewrite as a mixed or whole number:

EXAMPLE 5 Ethan ran $2\frac{1}{4}$ miles each day for 3 days. How far did Ethan run during the 3-day period?

Understand the problem

Create a plan

Find the answer

Check the answer

CONCEPT CHECK:

What is the first step for multiplying $1\frac{1}{2}$ and $1\frac{3}{4}$?

GUIDED EXAMPLES:

1. Multiply. $4\frac{2}{5} \cdot 3\frac{1}{2}$

2. Multiply. $5\frac{1}{4} \cdot 1\frac{3}{5}$

3. Multiply. $9 \cdot 4\frac{2}{3}$

4. Multiply. $2\frac{3}{8} \cdot 3\frac{2}{3}$

Name: _____   Date: _____

Instructor: _____   Section: _____

Notebook 4.4
Dividing Mixed Numbers

Write the steps for dividing mixed numbers:

1.

2.

3.

EXAMPLE 1 Divide. $1\dfrac{3}{4} \div 2\dfrac{4}{5}$

EXAMPLE 2 Divide. $3\dfrac{2}{5} \div 2\dfrac{2}{3}$

Write as improper fractions:

Invert the second fraction and multiply it by the first:

If improper, rewrite as a mixed or whole number:

EXAMPLE 3 Divide. $4 \div 3\dfrac{1}{3}$

Write as improper fractions:

Invert the second fraction and multiply it by the first:

If improper, rewrite as a mixed or whole number:

EXAMPLE 4 Divide. $3\dfrac{2}{5} \div \dfrac{1}{5}$

Write as improper fractions:

Invert the second fraction and multiply it by the first:

If improper, rewrite as a mixed or whole number:

OTHER NOTES

61

EXAMPLE 5 Megan purchased $12\frac{1}{4}$ yards of fabric. How many dresses can she make if it takes $2\frac{1}{3}$ yards to make each dress?

Understand the problem

Create a plan

Find the answer

Check the answer

CONCEPT CHECK:

Given the problem $\frac{9}{1} \div \frac{3}{4}$, what is the answer in simplest form?

GUIDED EXAMPLES:

1. Divide. $3\frac{5}{8} \div 5\frac{3}{4}$

2. Divide. $5\frac{1}{3} \div 2\frac{3}{8}$

3. Divide. $12\frac{1}{2} \div 5$

4. Marcus has $7\frac{1}{2}$ cups of flour. His cake recipe calls for $2\frac{1}{2}$ cups of flour. How many cakes can he bake?

Name: _____   Date: _____

Instructor: _____   Section: _____

Notebook 4.5
Adding Mixed Numbers

What must be true in order to add fractions?

What is the rule for adding like fractions?

EXAMPLE 1 Add. $3\dfrac{1}{3} + 2\dfrac{1}{3}$

Write the steps for adding mixed numbers:

1.

2.

3.

4.

EXAMPLE 2 Add. $7\dfrac{1}{3} + 4\dfrac{3}{5}$

Find the LCD:

EXAMPLES 3 & 4 Add.

$2\dfrac{5}{6} + 9\dfrac{2}{3}$

$2\dfrac{1}{3} + 15\dfrac{2}{3}$

EXAMPLES 5 & 6 Add.

$2 + 9\dfrac{2}{3}$

$2\dfrac{1}{6} + \dfrac{3}{4}$

EXAMPLE 5 Jordan bought $4\frac{2}{3}$ pounds of ham and $1\frac{1}{2}$ pounds of roast beef. How many pounds of lunch meat did Jordan buy?

Understand the problem

Create a plan

Find the answer

Check the answer

CONCEPT CHECK:

What is the first step in adding mixed numbers?

GUIDED EXAMPLES:

1. Add. $5\frac{4}{9}+3\frac{8}{9}$

2. Add. $35\frac{1}{2}+\frac{3}{10}$

3. Add. $24\frac{1}{2}+103\frac{1}{2}$

4. Aleta made a quilt using $1\frac{3}{4}$ yards of blue fabric and $3\frac{1}{2}$ yards of green fabric. How many yards of fabric did she use for the quilt?

Name: _____   Date: _____

Instructor: _____   Section: _____

Notebook 4.6
Subtracting Mixed Numbers

What must be true in order to subtract fractions?

What is the rule for subtracting like fractions?

EXAMPLE 1 Subtract. $5\frac{4}{7} - 2\frac{1}{7}$

Write the steps for subtracting mixed numbers:
1.

2.

3.

4.

Notes: When is borrowing needed?

What is another name for "borrowing"?

EXAMPLE 2 Subtract. $6\frac{1}{2} - 3\frac{1}{4}$

Is borrowing necessary?

EXAMPLE 3 Subtract. $9\frac{1}{3} - 2\frac{5}{8}$

Is borrowing necessary?

EXAMPLES 4 & 5 Subtract.

$$10\frac{2}{5} - 3$$

$$4 - 1\frac{1}{2}$$

EXAMPLE 5 James needs a piece of wood $4\frac{1}{3}$ feet long and buys a 8-foot board. If he cuts a piece $4\frac{1}{3}$ feet long, how long is the leftover piece of wood?

Understand the problem

Create a plan

Find the answer

Check the answer

CONCEPT CHECK: Consider the problem $4\frac{1}{6} - 2\frac{5}{6}$. What fraction should be added to the fraction $\frac{1}{6}$?

GUIDED EXAMPLES:
Subtract the following.

1. $4\frac{5}{9} - 2\frac{2}{9}$

2. $5\frac{1}{2} - 2\frac{5}{6}$

3. $17\frac{3}{5} - 8$

4. $7 - 2\frac{2}{9}$

Name: _____ Date: _____

Instructor: _____ Section: _____

Adding and Subtracting Mixed Numbers—Improper Fractions

This is an alternative method for adding and subtracting mixed numbers. This method involves rewriting all mixed numbers as

_____.

EXAMPLE 1 Add. $2\dfrac{4}{5}+1\dfrac{2}{5}$

EXAMPLE 2 Subtract. $3\dfrac{4}{5}-2\dfrac{2}{5}$

Write the steps for adding and subtracting mixed numbers using improper fractions:

1.

2.

3.

4.

5.

EXAMPLE 3 Add. $4\dfrac{2}{3}+3\dfrac{2}{3}$

EXAMPLE 4 Subtract. $5\dfrac{1}{8}-3\dfrac{3}{8}$

EXAMPLE 5 Add. $1\dfrac{2}{5}+3\dfrac{1}{2}$

EXAMPLE 6 Subtract. $5\dfrac{1}{3}-3\dfrac{5}{6}$

CONCEPT CHECK:

In the problem $6\dfrac{5}{6}-4\dfrac{1}{8}$, what is the LCD?

GUIDED EXAMPLES:

1. Subtract. $8\dfrac{2}{5}-3\dfrac{4}{5}$

2. Add. $4\dfrac{1}{9}+2\dfrac{5}{9}$

3. Add. $4\dfrac{1}{3}+3\dfrac{1}{4}$

4. Subtract. $5\dfrac{1}{2}-1\dfrac{1}{6}$

Name: _____   Date: _____

Instructor: _____   Section: _____

Notebook 4.8
Applications of Operations with Mixed Numbers

Write the steps for multiplying mixed numbers
1.
2.
3.

Write the steps for dividing mixed numbers.
1.
2.
3.

EXAMPLE 1 Divide $6\frac{2}{3} \div 1\frac{3}{5}$. Write your answer in simplest form.

Write the steps for adding mixed numbers.
1.
2.
3.
4.

Write the steps for subtracting mixed numbers.
1.
2.
3.
4.

EXAMPLE 2 Subtract $9\frac{2}{5} - 6\frac{3}{7}$. Write your answer in simplest form.

EXAMPLE 3 Kolton pitched $4\frac{2}{3}$ innings in game 1 of the district tournament.

He then pitched $6\frac{2}{3}$ innings in the next game. How many total innings did

Kolton pitch in these two games?
Understand the problem
Create a plan
Find the answer
Check the answer

69

EXAMPLE 4 It takes Max approximately $4\frac{1}{2}$ days to paint the inside of an average-sized house. About how many such houses can he paint during a 23-day pay period?

Understand the problem

Create a plan

Find the answer

Check the answer

EXAMPLE 5 Kimber can crochet $1\frac{1}{3}$ scarves from one spool of yarn. How many spools of yarn will she need in order to crochet $8\frac{1}{2}$ scarves?

Understand the problem

Create a plan

Find the answer

Check the answer

EXAMPLE 6 On Thanksgiving Day, the 5K Turkey Trot is run. The length of this race works out to around $3\frac{1}{10}$ miles. Marty runs the first $2\frac{1}{2}$ miles in 24 minutes. How much farther does she have to run to finish the race?

Understand the problem

Create a plan

Find the answer

Check the answer

CONCEPT CHECK: Heath is jogging $4\frac{3}{5}$ of a mile and has $1\frac{2}{3}$ miles left. How far has he run? Will this calculation require finding an LCD. Why or why not?

GUIDED EXAMPLES:

1. During registration, the advising center will be open for $7\frac{1}{2}$ hours with one advisor on duty. How many advisors will be needed if each advisor works a $1\frac{1}{2}$-hour shift?

2. The recipe for making one gallon of sweet tea calls $1\frac{1}{3}$ cups of sugar. How much sugar is needed to make $4\frac{1}{2}$ gallons of sweet tea?

3. Raekwon spent $15\frac{3}{4}$ hours volunteering at the local food pantry. If his goal is to volunteer 35 hours this semester, how many more hours does he need to spend?

70

Name: _____   Date: _____

Instructor: _____   Section: _____

Notebook 5.1
Decimal Notation

What are the three parts of a decimal number?

-
-
-

EXAMPLE 1 Identify the three parts of the decimal number 4.75.

What does the decimal part of a decimal number represent?
Then give some examples.

EXAMPLE 2 For each figure, write the decimal fraction and the decimal number that represents each shaded part.

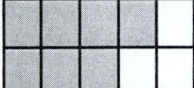

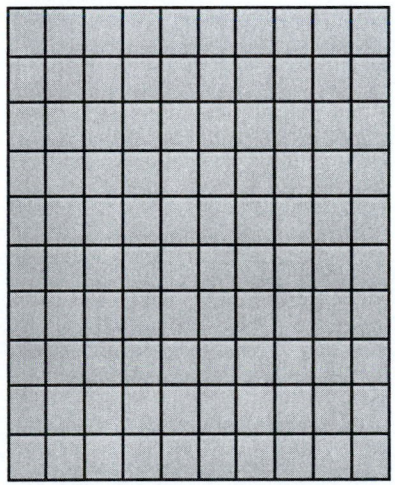

Whole Numbers				.	Decimals				
One Thousands	Hundreds	Tens	Ones	Decimal point	Tenths	Hundredths	One-Thousandths	Ten-Thousandths	Hundred-Thousandths

71

EXAMPLE 3 Read the decimals.

2.4

214.79

EXAMPLES 4–6 Read the following decimals.

2.3

4.17

12.0457

What are the steps for writing decimal in words?

1. Write the _____ _____ in words.
2. Write the word _____ for the decimal point.
3. Write the _____ _____ of the number in words.
4. Follow this by the _____ _____ of the last _____.

EXAMPLES 7 – 9 Write each decimal in words.

0.32

15.703

2.25

How do you write decimals in standard form?

1. Write the whole number in _____ _____.
2. Write the _____ in place of _____.
3. Use the given place value to determine

_____.

4. Write the decimal part in _____ _____-.

EXAMPLES 11 & 12 Write each decimal in standard form.

twenty-one and two hundred thirty-seven thousandths

fourteen and eight hundredths

CONCEPT CHECK:

Amanda wrote a five hundred seventy-nine and eighteen thousandths as 578.18. This is incorrect. How can she correct it?

GUIDED EXAMPLES:

1. Write the fraction and decimal represented by the picture.

2. Write in words. 4.128

Notebook 5.2
Comparing Decimals

Which is larger, 0.3 or 0.7?

EXAMPLES 1 & 2 Of each pair of decimals, which is larger?
0.14 and 0.41

2.315 and 2.871

Define "less than".

Define "greater than".

Note: Which number does the symbol point to?

EXAMPLE 3 Which is larger, 0.6 or 0.59?

Rule: What does "adding zeros to the end of a decimal" do to a number?

EXAMPLE 4 0.5 is the same as 0.50.

CAUTION!! Does inserting zeros at the beginning of a
decimal part change its value? Give an example.

Write the steps for comparing decimals:
1.
2.
3.

EXAMPLES 5 & 6 Fill in the blank with < or > to make a true statement.
4.13 _____ 4.90

0.983 _____ 0.7822

EXAMPLE 7 Arrange these decimals in order from smallest to largest.
1.23, 1.31, 1.2, 1.0567

EXAMPLE 8 Order the scores from smallest to largest.
During the 2998 Olympic Games, the U. S. Women Gymnastics Team scored 46.875 on the vault, 47.975 on the uneven bars, 47.25 on the balance beam, and 44.425 on the floor exercises.

CONCEPT CHECK:
True/False 2.6 < 5.4

GUIDED EXAMPLES:

1. Fill in the blank with <, > or =.   0.46 _____ 0.45

2. Fill in the blank with <, > or =.   0.251 _____ 0.25

3. Fill in the blank with <, > or =.   0.94 _____ 0.9324

4. Order from smallest to largest. 6.1, 5.23, 5.2

Notebook 5.3
Rounding Decimals

OTHER NOTES

EXAMPLE 1 Rasheed buys a camera for $328.28. There is a sales tax, which calculates to $20.5175. The store needs to round this tax to the nearest cent, or hundredths place, before adding it to Rasheed's bill.

Whole Numbers				.	Decimals				
One Thousands	Hundreds	Tens	Ones	Decimal Point	Tenths	Hundredths	One-Thousandths	Ten-Thousandths	Hundred-Thousandths

EXAMPLES 2–4 Each of the following often involves rounding of decimals:
*

*

*

Write the steps for rounding decimals to a given place value.
1. Underline the digit in the …

2. Look at the digit to the _____ of the underlined digit.

3. If that digit is:
   - 4 or less, …

   - 5 or more, …

4. Leave off all digits …

EXAMPLE 5 Round 13.45812 to the nearest thousandth.
   What digit is underlined?
   What is the digit to the right?
   Round up or down?
   Which digits are left off?

EXAMPLE 6 Round 6.7481 to the nearest hundredth.

Note: When might it be necessary to change the digit to the left of the underlined digit?

EXAMPLE 7 Round 7.9561 to the nearest tenth.

CONCEPT CHECK:
The sales tax for an item is $3.125 and needs to be rounded to the nearest cent. What number needs to be underlined?

GUIDED EXAMPLES:
1. Round to the nearest tenth. 4.65

2. Round to the nearest hundredth. 19.697

3. Round to the nearest thousandth. 3.1282

4. Round to the nearest tenth. 123.571

Notebook 5.4
Adding and Subtracting Decimals

How is adding and subtracting decimals similar to adding and subtracting whole numbers and like fractions?

EXAMPLE 1 Jennifer spent $2.97 for a meal and $1.19 for dessert, including tax. How much did she spend in total?

What is the procedure for adding and subtracting decimals?
1. Write the numbers _____, …
2. Write all decimals with …
3. Add or subtract, starting …
4. Place the decimal point …

EXAMPLES 2 & 3 Add. Check by estimating.
$3.27 + 15.2$ $4 + 12.3 + 0.571$

EXAMPLES 4 & 5 Find the difference. Check by estimating.
$13.24 - 7$ $21.5 - 16.43$

What is perimeter of a triangle?

How do you find the perimeter of a triangle?

EXAMPLE 5 Find the perimeter of the triangle.

3.5 m   1.5 m
4.2 m

EXAMPLE 6 At Sandee's Malt Shop, a hamburger costs $3.95, a hot dog costs $2.49, French fries cost $1.59, and onion rings cost $1.89. How much more does a hamburger with French fries cost than a hot dog with onion rings?
Understand the problem

Create a plan

Find the answer

Check your answer

CONCEPT CHECK:
A triangle has sides of length 10.9 in., 8.2 in., and 8.1 in. What is the proper expression for finding its perimeter?

GUIDED EXAMPLES:
1. Add. $20.52 + 6.9 + 13.8$

2. Add. $9 + 6.501 + 43.2$

3. Subtract. $246.72 - 98.264$

4. Subtract. $62.2 - 51.98$

Notebook 5.5
Multiplying Decimals

Show why multiplying $\frac{3}{10}$ and $\frac{2}{10}$ is the same as multiplying 0.2 and 0.3.

Write the steps for multiplying decimals.

1.

2.

3.

EXAMPLES 1 & 2 Multiply.

3.45(6.1) 0.152×0.23

EXAMPLES 3 & 4 Multiply.

3.2×0.008 4 (2.5)

What is the procedure for multiplying a decimal by a power of 10?

What do you do if the number of zeros is more than the number
of original decimal places?

EXAMPLE 5 Multiply.

5.793×100

79

EXAMPLES 6 & 7 Multiply.

79.251×10

42.8×1000

EXAMPLE 8 Ian's new car travels 25.6 miles for each gallon of gas. How many miles can he travel on a full tank, which contains 12 gallons of gas?

Understand the problem

Create a plan

Find the answer

Check your answer

CONCEPT CHECK:

Nataliya jogs seven days a week along a running path that is 2.9 miles long. Her goal is to jog 20 miles a week. Does she achieve her goal by jogging this path every day?

GUIDED EXAMPLES:

1. Multiply. 1.5×3.1

2. Multiply. 0.035×0.08

3. Multiply. 68.157×100

4. Multiply. $9.63 \times 10,000$

Notebook 5.6
Dividing Decimals

What is the number being divided?

What is the number the dividend is being divided by?

What is the answer to a division problem?

Label each part of a division problem in three different ways:

What is the rule for dividing a decimal by a whole number?

EXAMPLE 1 Divide. $6.3 \div 3$

EXAMPLE 2 Simon, Penny, Ashley, and James went out to eat. The total bill was $36.24. If they split the bill evenly, how much did each person pay?

Understand the problem

Create a plan

Find the answer

Check the answer

EXAMPLE 3 Divide. $6.7 \div 10$

Write the rule for dividing a decimal by a power of 10.

What do you do if the number of zeros is more than the number of whole number places?

EXAMPLE 4 Divide. $2.9 \div 1000$

What happens when we divide 3.8 by 4?
When working with decimals, there are no _____.

What are the three possibilities when dividing decimals?
1.

2.

3.

EXAMPLE 5 Divide 6.1 by 3.

Write the steps for dividing a decimal by a decimal.
1.

2.

3.

4.

EXAMPLE 6 Divide. $0.0612 \div 0.12$

EXAMPLE 7 Divide. Round to the nearest hundredth. $1.219 \div 0.3$

CONCEPT CHECK:
In the problem $19.21\overline{)24.567}$, what does the dividend become after the divisor
has been changed to a whole number?

GUIDED EXAMPLES:
1. Divide. $8.64 \div 4$

2. Divide. $245.967 \div 1000$

3. Divide. Round to the nearest tenth. $1.4 \div 0.6$

4. Divide. $5.564 \div 4.28$

Notebook 5.7
Order of Operations and Decimals

Write down the steps for using the order of operations:

P:

E:

MD:

AS:

Remember, certain symbols can act like parentheses. What are they?

What are the two memory tips for remembering the order of operations?

EXAMPLES 1 & 2 Simplify.

$3.6 + (0.2 + 0.3)^2$

    P:

    E:

    MD:

    AS:

$(12.6 - 11.4)^2 \div (0.2)^3$

    P:

    E:

    MD:

    AS:

EXAMPLES 3 & 4 Simplify.

$10.5 - 3.2 + 7.2 \div 3.6$

    P:

    E:

    MD:

    AS:

$10 \div 5(0.1)^2$

    P:

    E:

    MD:

    AS:

Note: A fraction bar can also act like _____ .

EXAMPLE 5 Simplify. $\dfrac{10.1 + 2.1(3)}{5.05 + 3.15}$

P:

E:

MD:

AS:

CONCEPT CHECK:
What is the second calculation involved in evaluating $19.12 - 9.6 \div 1.2 \times 1.5$?

GUIDED EXAMPLES:

1. Simplify. $7.1 - (0.6)(0.5 \div 125)^2$

2. Simplify. $12.6 \div 3 + (1.8 + 0.4)^2$

3. Simplify. $2.1 + 8.4 \times 3.12 - 1.37$

84

Notebook 5.8
Converting Fractions to Decimals

Every fraction can be written as, or converted to, an

_____.

Write $\dfrac{57}{100}$ as an equivalent decimal:

EXAMPLES 1 – 3 Write each fraction as an equivalent decimal.

$$\dfrac{1}{10} \qquad\qquad \dfrac{87}{100} \qquad\qquad \dfrac{119}{1000}$$

Does the denominator have to be a power of 10 in order to write the fraction as a decimal?

What are some common fractions and their decimal equivalents?

$$\dfrac{1}{4} \qquad\qquad \dfrac{1}{2} \qquad\qquad \dfrac{3}{4}$$

How do you write a fraction as a decimal?

EXAMPLES 4 & 5 Write the decimal as a fraction in simplest form.

$$\dfrac{3}{5}$$

$$\dfrac{7}{8}$$

Caution! The _____ is being divided by the _____.

EXAMPLE 6 Convert the fraction to a decimal.

$$\dfrac{2}{3}$$

What do we mean by a repeating decimal? What does the bar represent?

EXAMPLE 7 Convert the fraction to a decimal.
Round to the nearest thousandth.

$$\frac{4}{11}$$

EXAMPLES 8 & 9 Convert the fraction or mixed number to a decimal.

$$2\frac{3}{4}$$

$$\frac{11}{4}$$

EXAMPLE 10 The lead-off batter for the Raiders baseball team got on base 31 out of 70 times at bat. Write this fraction as a decimal rounded to the nearest thousandth.

CONCEPT CHECK:

Which procedure can be used to convert $5\frac{3}{4}$ to a decimal?

GUIDED EXAMPLES:

1. Convert the fraction to a decimal. $\dfrac{4}{5}$

2. Convert the fraction to a decimal. $\dfrac{1}{4}$

3. Convert the fraction to a decimal. $\dfrac{5}{8}$

4. Convert the fraction to a decimal. Round to the nearest hundredth. $\dfrac{2}{7}$

Notebook 5.9
Converting Decimals to Fractions

OTHER NOTES

Write 0.9 as a fraction:

Write 0.23 as a fraction:

What are the steps for writing a decimal as a fraction or mixed number?
1.

2.

3.

4.

Note: What do you know about the number of decimal places and
the number of zeros in the power of 10 in denominator?

EXAMPLE 1 Write 5.277 as a fraction in simplest form.

EXAMPLES 2 & 3 Write the decimal as a fraction in simplest form.
0.96

0.0201

EXAMPLES 4 & 5 Write the decimal as a fraction in simplest form.
0.425

0.02001

EXAMPLES 6 & 7 Write the decimal as a fraction in simplest form.
4.0358

5.2

CONCEPT CHECK:
How many zeros will be in the denominator of 6.423 when written as a fraction?

GUIDED EXAMPLES:
Write as a fraction in simplest form.
1. 0.5

2. 0.59

3. 0.256

4. 10.1281

88

Name: _____   Date: _____

Instructor: _____   Section: _____

Notebook 6.1
Ratios

What is a ratio?

What are the different ways in which a ratio can be written?
Give examples.

EXAMPLE 1 A gallon of lemonade is made by mixing 2 cups of lemon juice with 14 cups of water. Write the ratio of lemon juice to water in a gallon of lemonade.

What is the rule for writing ratios as fractions?
    The quantity appearing first is …

    The quantity appearing second is …

EXAMPLE 2 Write 5 hours to 7 hours as a ratio. Use all three ways.

Do we need to include the units in ratios? _____ Why or why not?

EXAMPLES 3 & 4 Write the following ratios as fractions in simplest form.

    5 inches to 11 inches

    4 ounces: 32 ounces

EXAMPLES 5 & 6 Write the following ratios as fractions in simplest form.
$1.23 to $3.75

Joseph lost $5\frac{1}{2}$ pounds, while his friend Demarcus lost $3\frac{1}{4}$ pounds.

Find the ratio of the weight that Joseph lost to the weight that Demarcus lost.

EXAMPLE 7 A cake recipe requires $2\frac{1}{2}$ cups of flour and $1\frac{3}{4}$ cups of sugar.

Write the ratio of the amount of flour to the amount of sugar as a fraction in the simplest form.

Understand the problem

Create a plan

Find the answer

Check the answer

CONCEPT CHECK: True/False The quantity appearing second appears in the numerator.

GUIDED EXAMPLES:

Write each ratio as a fraction in simplest form.

1. 11 inches to 9 inches

2. $1\frac{1}{2}$ hours to $3\frac{1}{2}$ hours

3. $3\frac{3}{4}$ gallons to $4\frac{1}{2}$ gallons

4. Mark's truck can tow 2.5 tons and Kyle's truck can tow 5.0 tons. Find the ratio of the towing capacity of Mark's truck to the towing capacity of Kyle's truck.

90

Name: _____   Date: _____

Instructor: _____   Section: _____

Notebook 6.2
Rates

What is a rate?

What are two common examples of rates?

What is the rule for writing rates as fractions?
    The quantity appearing first is …

    The quantity appearing second is …

Do we need to include the units in rates? _____ Why or why not?

EXAMPLE 1 Write the rate as a fraction in simplest form.
    240 miles per 10 gallons

EXAMPLE 2 Write the rate as a fraction in simplest form.
    18 tomatoes for 4 pots of stew

EXAMPLES 3 & 4 Write the following ratios as fractions in simplest form.

    20 door prizes for 110 people

    $250 for 20 hours

What is a unit rate?

What is the difference between a rate and a unit rate?

EXAMPLE 5 Find the unit rate.

A 6-pack of juice bottles costs $3.00. Find the unit rate of cost per 1 bottle.

Note: To find a rate, write the rate as …

To find a unit rate, write the rate as …

What is the procedure for finding the better buy?

EXAMPLE 6 Which of the following is a better buy?
A 12-ounce box of Antonio's Macaroni costs $2.08.

A 15-ounce box of Sally's Macaroni costs $3.05.

Solution: Find the cost per ounce for each box:

What is the cost per ounce for a 12-ounce box?

What is the cost per ounce for a 15-ounce box?

Which one is the better buy?

CONCEPT CHECK: True/False The best buy has the highest unit cost.

GUIDED EXAMPLES:
Write the rate as a fraction in simplest form.
1. 15 departments for 10 salespeople

2. $210 for 28 pizzas

3. Find the unit rate of miles per gallon. 118 miles for 5 gallons

4. Murphy's Crunch cereal comes in a 16-ounce box for $3.20 and Dawn's Nuggets cereal comes in a 25-ounce box for $6.00. Which is the better buy?

Name: _____   Date: _____

Instructor: _____   Section: _____

Notebook 6.3
Proportions

What is an equation?

What is a proportion?

EXAMPLES 1 & 2 For each example, write as a proportion.

    3 is to 6 as 12 is to 24

    Caleb paid \$60 for 4 tickets and Beth paid \$90 for 6 tickets.

CAUTION!! Is consistency important when setting up a proportion? What does that mean?

What is the procedure for determining if a statement is a proportion?

    If the cross products are equal, then the statement is …
    If the cross products are not equal, then the statement is …

EXAMPLES 3 & 4 Determine if the statement is a proportion.

$$\frac{6}{10} = \frac{10}{35}$$

$$\frac{2}{3.1} = \frac{3}{3.7}$$

If we know three of the four numbers in a proportion, $\frac{a}{b} = \frac{c}{d}$, such that $b \neq 0$ and $d \neq 0$, can we find the fourth number?
Does it matter which one is missing?

EXAMPLE 5 Find the missing number in the proportion.

$$\frac{3}{9} = \frac{4}{?}$$

Write the steps for finding missing number in a proportion:

1. Calculate the _____ and set them

   _____.

2. Divide both sides of the equation by …

EXAMPLE 6 Find the missing number in the proportion.

$$\frac{5}{6} = \frac{n}{12}$$

Check:

EXAMPLE 7 The proper dosage of a children's medicine is 40mg for every 50 pounds. How much medicine do you give a child weighing 75 pounds?
Understand the problem

Create a plan

Find the answer

Check the answer

CONCEPT CHECK: What symbol does ALL equations contain?

GUIDED EXAMPLES:

1. Write the following as a proportion.

   7 is to 12 as 14 is to 24

2. Determine if the statement is a proportion.

   $$\frac{3.6}{4.8} \stackrel{?}{=} \frac{2.4}{3.2}.$$

3. Find the missing number In the proportion.

   $$\frac{5}{7} = \frac{n}{28}$$

4. You can drive 200 miles on 8.5 gallons of gas. How much gas will it take to drive 600 miles?

Notebook 6.4
Percent Notation

What does percent mean?

OTHER NOTES

What are some examples of percents?

How do you write a percent as a fraction?

EXAMPLES 1–3 Write each percent as a fraction in simplest form.

47%

14%

136%

When is it easy to convert fractions to percents?

EXAMPLES 4 & 5 Write each fraction as a percent.

$\dfrac{7}{10}$

$\dfrac{61}{100}$

EXAMPLES 6–8 Answer the following questions involving percents.
Penny answered 87 out of 100 problems correctly on her math test.
What percent of the problems did she answer correctly?

68 out of 100 cars in a parking lot are white. What percent of the cars are white?

46 out of 100 children prefer strawberry ice cream. What percent of children do <u>NOT</u> prefer strawberry ice cream?

What does 100% represent?

What does more than 100% mean?

What does less than 100% mean?

EXAMPLES 9–12 Are the following common percents less than, equal to, or greater than one whole unit?

50%	225%
100%	150%

What is the procedure for finding the percent of a number using fractions?
To find $p\%$ of a number, …

Find 4% of 200.

EXAMPLE 13 A company produced 2500 personal computers, and 3% of them were defective. How many of the computers were defective?

Understand the problem

Create a plan

Find the answer

Check the answer

CONCEPT CHECK: Which percent below represents more than one whole unit?
99%, 2%, 101% or 50% ?

GUIDED EXAMPLES:

1. Write 62% as a fraction in simplest form.

2. Write $\dfrac{83}{100}$ as a percent.

3. Jackie has 100 marbles and 29 of them are red. What percent of the marbles are red?

4. A company produced 5000 digital cameras and 1% of them were defective. How many of the cameras were defective?

Notebook 6.5
Percent and Decimal Conversions

Recall: What does percent mean?

EXAMPLE 1 Write 51% as a decimal.

What is the procedure for writing a percent as a decimal?

Note: What do you do if there is no decimal point in the original percent?

CAUTION!! It may be necessary to insert zeros in order to …

EXAMPLES 2 – 5 Write each percent as a decimal.

135%

24.9%

$74\frac{1}{2}\%$

4.5%

EXAMPLE 6 Write 0.71 as a percent.

What is the procedure for writing a decimal as a percent?

CAUTION!! It may be necessary to add zeros in order to …

EXAMPLES 7–10 Write each decimal as a percent.

2.3

0.095

0.75

0.004

OTHER NOTES

97

What is the procedure for finding the percent of a number?

For example, find 10% of 500.

EXAMPLE 11 A company made 7500 tables, and 20% of these tables were made from oak. How many of the tables were made from oak?

Understand the problem

Create a plan

Find the answer

Check the answer

CONCEPT CHECK: Carrie converted 0.53 to a percent and got an answer of 0.0053%. What was her mistake?

GUIDED EXAMPLES:

Write the percent as a decimal.

1. $9\dfrac{3}{4}\%$

2. $87\dfrac{1}{4}\%$

Write the decimal as a percent.

3. 0.225

4. A company made 6000 folders, and 5% of the folders were red. How many of the folders were red?

Notebook 6.6
Percent and Fraction Conversions

Recall: What does $p\%$ mean?

Recall: What are two ways you can convert $p\%$ to a fraction?

    1.

    2.

Note: Dividing p by _____ and multiplying p by _____ both result in the same answer.

EXAMPLE 1 Write 8% as a fraction in simplest form.

EXAMPLES 2–4 Write each percent as a fraction in simplest form.

    Joe sells cars and receives $3\frac{1}{4}\%$ on each sale.

    Property tax is $\frac{3}{10}\%$ off the estimated value.

    The sales tax is 4.5% of the purchase price.

How do you write a fraction as a percent?
If the denominator is 100:

If the denominator is not 100:

EXAMPLES 5 & 6 Write each fraction as a percent.

    $\dfrac{32}{100}$

    $\dfrac{5}{8}$

EXAMPLE 7 Write as a percent. Round to the nearest tenth of a percent.

    $\dfrac{4}{7}$

OTHER NOTES

Note: Rounding to the nearest tenth of a percent requires performing the division to _____ decimal places.

CAUTION!! When is rounding performed during a problem?

Summary:

Fraction	Decimal	Percent
$\dfrac{32}{100}$		
$\dfrac{5}{8}$		
$\dfrac{4}{7}$		

CONCEPT CHECK: If you are asked to solve a problem with the directions "write $\dfrac{6}{11}$ as a percent, rounded to the nearest tenth of a percent." What is the result before you round?

GUIDED EXAMPLES:
Write the % as a fraction in simplest form.
1. 65%

2. $\dfrac{1}{4}\%$

Write the fraction as a percent. Round to the nearest tenth of a percent if necessary.
3. $\dfrac{3}{8}$

Name: _____   Date: _____

Instructor: _____   Section: _____

Notebook 6.7
The Percent Equation

What is the Percent Equation?

What does it look like in equation form?

"What amount is   a percent   of   the whole?

Notes:

Usually "base" appears ….

"is" translates to …

"of" translates to …

Any of the quantities, _____, _____, or _____,
can be the unknown quantity

Usually a is used for _____, p for the _____,
and b for _____.

In symbols, the percent equation is _____.

EXAMPLES 1 & 2 Finding the amount.
What amount is 30% of 140?

60% of 90 is what amount?

EXAMPLES 3 & 4 Finding the base.
34 is 50% of what number?

60% of what number is 42?

EXAMPLE 5 45 is what percent of 180?

EXAMPLE 6 What percent of 30 is 10? Round to the nearest tenth of a percent.

Note: Rounding to the nearest tenth of a percent requires performing the division to _____ decimal places.

CAUTION!! True/False Rounding can be done at any time.

CONCEPT CHECK: For the problem "45 is 50 percent of what number," which quantity is unknown?

GUIDED EXAMPLES:
Use the percent equation to find the following. Round to the nearest tenth of a percent as needed.
1. What amount is 75% of 240?

2. 90 is 15% of what number?

3. 18 is what percent of 60?

4. 20 is what percent of 120?

Notebook 6.8
The Percent Proportion

OTHER NOTES

What does the percent proportion state?

What is the percent proportion?
    Usually "base" appears ….

        Any of the quantities, _____, _____, or _____,
        can be the unknown quantity

        Usually a is used for _____, p for the _____,
        and b for _____.

        Since p is divided by 100, do we need to convert from a decimal?

Note: The percent proportion is an alternative to _____.

EXAMPLE 1 What amount is 20% of 85?
    Percent =

    Base =

    Find the amount:

EXAMPLE 2 21 is 10.5% of what number?
    Percent =

    Amount =

    Find the base:

EXAMPLE 3 83 is what percent of 332?
    Amount =

    Base =

    Find the percent:

EXAMPLE 4 4 is what percent of 48? Round to the nearest tenth of a percent.

Amount =

Base =

Find the percent:

CONCEPT CHECK: What is the unknown quantity in the problem "19 is 17% of what number"?

GUIDED EXAMPLES:

1. What amount is 25% of 64?

2. 25 is 12.5% of what number?

3. 12 is what percent of 60?

4. 30 is what percent of 45? Round to the nearest tenth of a percent.

Notebook 6.9
Percent Applications

Write the Percent Equation:

Write the Percent Proportion:

Note: Either the percent equation or the percent proportion can be used to solve problems involving _____.

Note: If you know the percent and the base, which one would be better to use?

Write the rule for finding the amount of sales tax and total cost.
Amount of Sales Tax =

Total Cost =

EXAMPLE 1 Eli purchased a wrist watch for $60. If the sales tax rate is 7%, how much sales tax did Eli pay? What was the total cost of the watch?
       Amt of sales tax:

       Total cost:

Write down the rule for finding the amount of discount and sale price.
Amount of Discount =

Sale Price =

EXAMPLE 2 Lucky's Shoes is having a 30% discount sale. How much will Jan save on a $40 pair of shoes? What is the sale price of the shoes?
       Amt of discount:

       Sale price:

Write the rule for finding the percent of change.

Percent of change =

EXAMPLE 3 Enrollment at the local high school went from 320 students to 480 students in one year? Find the percent of increase.

      Amt of increase:

      Percent change:

What do you know about simple interest?

      It is often used in…

      It uses an annual rate to calculate ….

      It is usually calculated _____, not _____.

Write down the rule for finding simple interest and total amount paid.

Amount of interest =

Total amount paid =

Notes:

      Principal is the amount of the _____.

      The interest rate is the percent charged per year, expressed as a _____.

      The amount of time must be expressed in _____.

EXAMPLE 4 Calculate the amount of interest and the total amount paid for a loan of $2000 for 3 years at a 4% annual rate of interest.

      Interest:

      Total amount paid:

CONCEPT CHECK: Arty is buying a $50 watch. Sales tax is 9.25%. What is his first step to calculate the total cost of the watch?

GUIDED EXAMPLES:

Alesha purchases a 3-speed bike for $450.

1. If the sales tax is 8%, how much sales tax did Alesha pay?

2. What was the total cost of his bike?

James bought a washer and dryer combination for $1100. He bought it on credit for 1 year at 6% simple interest.

3. Find the amount of interest James will pay.

4. Find the total cost of the washer and dryer, including interest.

Notebook 7.1
U.S. Length

Length can be measured in various units.
What is one way to measure length?

What determines the unit of length you use?

TABLE 1 U.S. Units of Length	
Unit	**Abbreviation**
inch	
foot	
yard	
mile	

TABLE 2 U.S. Length Conversions
12 in. = 1 ft
36 in. = 1 yd
3 ft = 1 yd
5280 ft = 1 mi
1760 yd = 1mi

_____ are used to convert units of measure.

What is a unit fraction?
Give some examples of unit fractions for U.S. length:

What is the rule for finding the appropriate unit fraction?

Write the steps for converting using unit fractions.
1.
2.
3.

EXAMPLE 1 Convert the following using unit fractions. 24 yards to feet

EXAMPLE 2 Convert the following using unit fractions. 15,840 feet to miles

EXAMPLE 3 Convert the following using unit fractions. 4.5 yards to inches

EXAMPLE 4 Convert the following using unit fractions. 90 inches to yards

107

Write the steps for converting to mixed units.

1.

2.

3.

EXAMPLE 5 Gage is 54 inches tall. Convert this to feet and inches.

CONCEPT CHECK:

When converting 56 inches to feet and inches, what number will represent the number of inches in the final answer?

GUIDED EXAMPLES:

1. Convert using unit fractions. 2.4 miles to yd

2. Convert using unit fractions. 144 in. to yd

3. Convert 102 in. to ft.

4. Convert the height to feet and inches. J.C. is 75 inches tall.

Notebook 7.2
U.S. Weight and Capacity

What is the difference between weight and mass?
        Weight is …

        Mass …

Which is used in the U.S. System of Measurement?
Which is used in the Metric System of Measurement?

What determines which unit of weight you use?

TABLE 1 U.S. Units of Weight	
Unit	**Abbreviation**
ounce	
pound	
ton	

TABLE 2 U.S. Weight Conversions
16 oz = 1 lb
2000 lb = 1 ton

Give some examples of unit fractions for U.S. units of weight.

What is the rule for finding the appropriate unit fraction?

Write the steps for converting using unit fractions.
1.
2.
3.

EXAMPLE 1 Convert the following using unit fractions. 7.5 tons to pounds

EXAMPLE 2 Convert the following using unit fractions. 64 ounces to pounds

Give some examples of unit fractions for U.S. units of capacity.

109

EXAMPLE 3 Convert the following using unit fractions. 7.5 pints to cups

EXAMPLE 4 Convert the following using unit fractions. 26 quarts to gallons

Write the steps for converting to mixed units.
1.

2.

3.

EXAMPLE 5 LaShonda buys a bottle of ketchup that contains 44 fl oz.
Convert this to pints and fluid ounces.

CONCEPT CHECK:
Which unit would be the most appropriate to measure the weight of a cement
truck?

GUIDED EXAMPLES:
1. Convert 5000 pounds to tons.

2. Convert 9 lb to oz.

3. Convert 52 qt to gallons.

4. Convert 38 fluid ounces to _____ pints _____ ounces.

110

Name: _____   Date: _____

Instructor: _____   Section: _____

Notebook 7.3
Metric Length

In the United States, what units of measurement do we use for length?

In the metric system, what is the basic unit of length?
All metric units of length are based on the _____.

How long is a meter?

Which metric units are shorter than the meter?
Which are longer than the meter?

Metric Units of Length references:
Which unit is about as thick as a larger paper clip wire?
Which unit is about the width of your pinky finger?
Which unit is about the width of your hand?
Which unit is slightly longer than a yard?
Which unit is a little more than 30 feet?
Which unit is about the length of a football field?
Which unit is about 0.6 of a mile?

What does each metric prefix mean?
milli: centi: deci:

deka: hecto: kilo:

Recall: A unit fraction is a fraction whose value is _____.

Write the rule of finding the appropriate unit fraction:

EXAMPLE 1 Convert the following. 400 centimeters to meters

What phrase can help you to remember the order of the metric prefixes?

Note: The larger units are on the _____; the smaller units on the _____.
When converting larger units to smaller units, which direction do you move?
When converting smaller units to larger units, which direction do you move?

EXAMPLES 2 & 3 Convert the following.

87.45 millimeters to decimeters

    How many places do you move?

    Which direction do you move?

96.3 kilometers to meters

    How many places do you move?

    Which direction do you move?

EXAMPLE 3 Convert the following. 150 dekameters to centimeters

    How many places do you move?

    Which direction do you move?

CONCEPT CHECK:

When converting 2 dekameters to millimeters, how many places should the decimal point be moved and in which direction?

GUIDED EXAMPLES:

1. Convert 45.1 m to km.

2. Convert 3.1 hm to cm.

3. Convert 46 cm to mm.

4. Convert 25 cm to meters.

Name: _____   Date: _____

Instructor: _____   Section: _____

In the United States, what units of measurement do we use for weight?

In the metric system, what is the basic unit of mass?
All metric units of mass are based on the _____.

How much is a gram?

How much is a kilogram?

Which metric units are smaller than the gram?
Which are larger than the gram?

Recall each metric prefix:
milli: centi: deci:

deka: hecto: kilo:

Recall: A unit fraction is a fraction whose value is _____.

EXAMPLE 1 Convert the following. 32 centigrams to grams

What phrase can help you to remember the order of the metric prefixes?

Note: The decimal point in the number moves the _____
of places and in the _____ as the number of moves
in the list.

EXAMPLES 2 & 3 Convert the following.
123.5 milligrams to grams
  How many places do you move?
  Which direction do you move?

216 kilograms to centigrams
  How many places do you move?
  Which direction do you move?

OTHER NOTES

113

In the United States, what units of measurement do we use to measure liquid capacity?

In the metric system, what is the basic unit of capacity?
All metric units of capacity are based on the _____.

How much is a liter?

How much is a kiloliter?

Which metric units are smaller than the liter?
Which are greater than the liter?

How much is one milliliter?

EXAMPLES 4 & 5 Convert the following.
900 milliliters to liters
    How many places do you move?
    Which direction do you move?

83.2 liters to centiliters
    How many places do you move?
    Which direction do you move?

CONCEPT CHECK:
Which is the most appropriate unit to measure the capacity of a pool?

GUIDED EXAMPLES:
1. Convert 4.08 mg to dg.

2. Convert 9.332 kg to cg.

3. Convert 4.03 mL to cL.

4. Convert 5.2 L to cc.

Name: _____   Date: _____

Instructor: _____   Section: _____

Notebook 7.5
Converting Between U.S. and Metric Units

Write down the unit fractions for converting lengths.
   U.S. to Metric (Length)

Recall the fraction used to make conversions:

   Metric to U.S. (Length)

EXAMPLE 1 Convert the following. 44.02 miles to kilometers
   Conversion:

EXAMPLE 2 The bases in a major league baseball stadium are 90 feet apart.
In terms of meters, how far apart are the bases?
   Conversion:

Write down the unit fractions for converting mass.
   U.S to Metric (Weight/mass)

   Metric to U.S. (Weight/mass)

EXAMPLE 3 Convert the following. 5.2 kilograms to pounds
   Conversion:

EXAMPLE 4 Convert the following. 6 pounds to grams
  Conversion:

Write down the unit fractions for converting capacity.
  U.S to Metric (Capacity)

  Metric to U.S. (Capacity)

EXAMPLE 5 Convert the following. 6 pounds to grams
  Conversion:

EXAMPLE 6 Convert the following. 3 quarts to liters
  Conversion:

CONCEPT CHECK:
A recipe lists the number of grams needed for an ingredient, but Sarah only has
a scale that uses U.S. measurements. She will most likely convert the amount
from the recipe to which unit?

GUIDED EXAMPLES:
1. Convert 225.6 m to yd.

2. Convert 2.6 in. to yd. Round to the nearest hundredth.

3. Convert 12.6 gal to L.

4. Convert 16.45 L to gal. Round to the nearest tenth.

Notebook 7.6
Time and Temperature

Write down the unit fractions for converting times.

Recall the fraction used to make conversions:

EXAMPLE 1 Convert 195 minutes to hours.

EXAMPLE 2 Convert 4.4 hours to seconds.

What are the two scales for temperature?

What is the basic unit for each scale?

What are some common temperature equivalencies?
   Water freezes:
   Room temperature:
   Water boils:

What is the formula used to convert from Fahrenheit to Celsius?

EXAMPLE 3 Convert 36° Fahrenheit to Celsius. Round to the nearest tenth.

OTHER NOTES

What formula is used to convert from Celsius to Fahrenheit?

EXAMPLE 4 Convert 81° Celsius to Fahrenheit. Round to the nearest tenth.

EXAMPLE 5 Marquez, who lives in the United States, purchased a heat pump that was made in Belgium. The heat pump can operate efficiently up to a temperature of 35° Celsius. To the nearest degree, what is this temperature in Fahrenheit?
Understand the problem

Create a plan

Find the answer

Check the answer

CONCEPT CHECK:

What is the unit fraction needed to convert seconds to minutes?

GUIDED EXAMPLES:
1. Convert 5 hours to minutes using unit fractions.

2. Convert 600 hours to days.

3. Convert 60° F to C.

4. Convert 10.3° C to F.

Name: _____   Date: _____

Instructor: _____   Section: _____

Notebook 8.1
Lines and Angles

What is a point?
How is a point represented?
Ex.

What is a line?
How is a line named?
Ex.

What is a line segment?
How is a line segment named?
Ex.

What is a ray?
How is a ray named?
Ex.

What is an angle?
How is an angle named?
Ex.

EXAMPLES 1–3 Identify each of the following figures as a line,
a line segment, or a ray. Give the name of each.

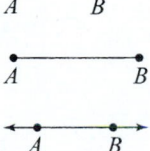

How is an angle formed?
An angle is measured by _____.
How is an opening measured?

Define four types of angles and give an example of each.
1.

2.

3.

4.

119

EXAMPLES 4 & 5 Identify each angle below as straight or right.

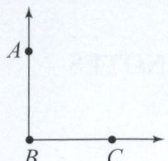

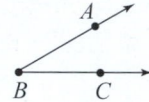

Identify each angle below as either acute or obtuse.

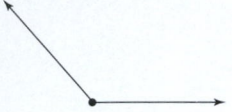

Two angles are _____ if the sum of their measures is _____.
Give an example.

Two angles are _____ if the sum of their measure is _____.
Give an example.

EXAMPLES 6 & 7 Identify each angle pair as complementary, supplementary, or neither. You will need to draw the angles.

EX.   EX.

Find the complement of a 65° angle.

Find the supplement of a 75° angle.

CONCEPT CHECK:
How do you find the complement of an angle?

GUIDED EXAMPLES:
1. Identify the figure as a line, a line segment, or a ray.

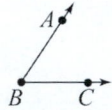

2. Identify the angle as acute, obtuse, or neither.

3. Identify the angle pair as complementary, supplementary, or neither.
   angles of 54° and 25°

4. Find the measure of the complement of a 23° angle.

120

Name: _____ Date: _____

Instructor: _____ Section: _____

Notebook 8.2
Figures

What is a polygon?

What does it mean that a line is one-dimensional?

Why are polygons considered two-dimensional?

A figure can be classified by the number of _____.
A 3 sided figure is a _____.

A 4 sided figure is a _____.

TRIANGLES–Draw an example of each.
What kind of triangle has all three sides equal in length?

What kind of triangle has two sides equal in length?

What kind of triangle has all three sides of different lengths?

Property: The sum of the measures of the three angles of a triangle is _____.

A triangle can also be classified according to the _____.
A triangle with a 90 angle is a _____.

A triangle with three acute angles is an _____.

A triangle with an obtuse angle is an _____.

EXAMPLE 1 If two angles of a triangle measure 55° and 70°, find the
measure of the third angle. Then, identify the triangle as acute, right, or obtuse.

QUADRILATERALS–Draw an example of each.
Rectangle   Square   Trapezoid

Parallelogram   Diamond   Kite

121

A _____ is a quadrilateral with opposite sides that are _____ in length and four angles that are _____.
Ex.

A square is a quadrilateral with _____ that are _____ in length and_____ angles that are _____.
Ex.

What are parallel lines?
What is the symbol for parallel lines?
Ex.

A parallelogram is a _____, or four sided figure, with opposite sides that are _____ and _____.

What is the difference between a rectangle and a parallelogram?

A _____ is a quadrilateral with only _____ pair of _____ that are parallel.

CONCEPT CHECK:
A triangle has angles that measure 57°, 72°, and 51°. What kind of triangle is it?

GUIDED EXAMPLES:
1. Find the measure of the third angle in each triangle.
   triangle with angles of 27° and 32°

2. Identify the triangle as acute, right, or obtuse.
   a triangle with angles of 55°, 75°, and 50°

3. Identify each triangle as acute, right, or obtuse.
   a triangle with angles of 45°, 45°, and 90°

4. Identify the figure below.

Name: _____   Date: _____

Instructor: _____   Section: _____

Notebook 8.3
Perimeter—Definitions and Units

What is a polygon?

Define perimeter.

How do you find the perimeter of a polygon?

Perimeter always represents _____ and is measured in units of
_____, such as …

Which of these situations, if any, would involve perimeter?
　　Painting a room
　　Spreading grass seeds
　　Hanging gutters around a house

Which of these measurements, if any, could represent perimeter?
　　19 inches
　　34 cups
　　2.3 square feet

EXAMPLE 1 Find the perimeter of the figure.
You will need to draw the triangle.

EXAMPLE 2 Find the perimeter of the figure.
You will need to draw the figure.

EXAMPLE 3 Find the perimeter of the figure.
You will need to draw the figure.

EXAMPLE 4 Find the distance around the infield of a major league baseball field. You will need to draw the figure.

CONCEPT CHECK:
A polygon is a closed figure with _____ or more sides.

GUIDED EXAMPLES:
1. Find the perimeter.

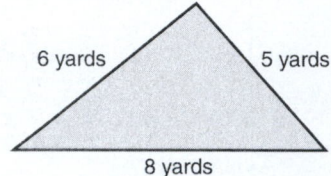

2. Find the perimeter.

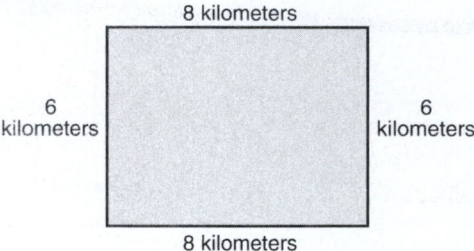

3. Tennessee is shaped like a parallelogram. Find the approximate length of the border around the state using the following figure.

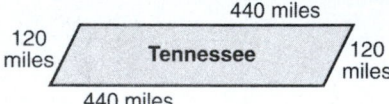

4. Find the perimeter of the figure.

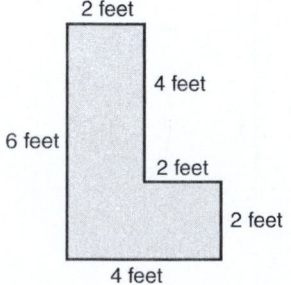

124

Name: _____   Date: _____

Instructor: _____   Section: _____

Notebook 8.4
Finding Perimeter

To find the perimeter of a figure, _____.
Perimeter is measured in units of _____.
A _____ is a three-sided geometric figure.
A _____ is a quadrilateral with opposite sides that
are _____ in length and four angles that are _____.
A _____ is a quadrilateral with four sides that are
_____ in length and four angles that are _____.

This video focuses on learning _____ for finding _____.

What is the formula for finding the perimeter of a triangle?
Ex.

EXAMPLE 1 Use the formula for the perimeter of a triangle to find the
perimeter of a triangle with side lengths of 3.5 cm, 2.5 cm, and 4 cm.

What is the formula for finding the perimeter of a rectangle?
Ex.

EXAMPLE 2 Use the formula for the perimeter of a rectangle to find the
perimeter of a rectangular field with a length of 40 feet and a width of 30 feet.

What is the formula for finding the perimeter of a square?
Ex.

EXAMPLE 3 Find the perimeter of a square to find the perimeter of a square
court with side lengths of 14 meters.

EXAMPLES 4–6 For each figure, find the perimeter.

A triangular garden with side lengths of 5 feet, 7 feet and 10 feet.

A rectangular photo with length 20.5 cm and a width 10 cm.

A square tabletop with side lengths of $2\frac{1}{2}$ m.

CONCEPT CHECK:

To find the perimeter of a triangle, how many side lengths must be known?

GUIDED EXAMPLES:

Use the appropriate perimeter formula to find the perimeter of the figures shown.

1.

7 yards

9 yards

6 yards

2.

9 inches

6 inches

3.

17.2 cm

Notebook 8.5
Area—Definitions and Units

How do you find the area of a figure?

Area is measured in _____, such as …

What does one square unit look like?

  Square inch:

  Square centimeter:

Which of these situations, if any, would involve area?
  Laying a carpet
  Putting up a fence
  Filling a swimming pool

Which of these units, if any, could represent area?
  36 inches
  6 gallons
  14 square feet

EXAMPLES 1 & 2 How many square units are needed to cover each figure completely? You will need to draw the figures.

OTHER NOTES

EXAMPLE 3 Draw two different rectangles, each of which has an area of 6 square units.

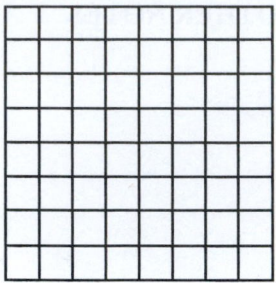

EXAMPLE 4 Find the area of a square that measures 5 cm on each side.

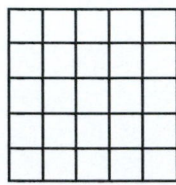

EXAMPLE 5 Find the area of a right triangle with a base and height of 4 in.

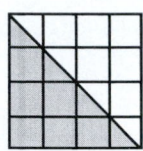

CONCEPT CHECK: A triangle covers 9 whole squares and 5 half squares. What would this area be?

GUIDED EXAMPLES:

1. Find the area of a right triangle with a base and height of 3 in.

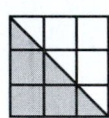

2. Draw a rectangle which has an area of 12 square units.

3. Which of these situations, if any, would involve area?
   a. Measuring flour for a cake   c. Framing a picture
   b. Painting a room   d. Laying a tile floor

Name: _____ Date: _____

Instructor: _____ Section: _____

Notebook 8.6
Finding Area

Area is the measure of the _____ inside a figure.
To find the area, _____.
Area is measured in _____.
A _____ is a three-sided geometric figure.
A _____ is a quadrilateral with opposite sides that
  are _____ in length and four angles that are _____.
A _____ is a quadrilateral with four sides that are
_____ in length and four angles that are _____.

This video focuses on learning _____ for finding _____.

What is the formula for finding the area of a rectangle?
Ex.

EXAMPLE 1 Use the formula for the area of a rectangle to find the area of the
photo. You will need to draw the rectangle.

What is the formula for finding the area of a square?
Ex.

EXAMPLE 2 Use the formula for the area of a square to find the area of a
square deck with side lengths of 3 meters.

What is the formula for finding the perimeter of a triangle?
How does the height vary with different types of triangles?
Ex.

EXAMPLE 3 Use the formula for the area of a triangle to find the area of a
right triangle with a base of 4 cm and a height of 3 cm.

OTHER NOTES

129

EXAMPLE 4 Find the area of the infield of a major league baseball diamond, which is a square whose sides are 90 feet long.
You will need to draw the diagram.

CONCEPT CHECK:
An 8 in. by 11 in. piece of graph paper has a grid with lines that are one inch apart. In other words, it is covered in squares that measure 1 inch on each side. How many squares, or square units, cover the graph paper?

GUIDED EXAMPLES:
Find the area of the following.
1. A rectangle whose length is 15 inches and width is 4 inches.

2. A square whose sides are 6 cm long.

3. A right triangle with a base of 6 cm and a height of 5 cm.

4. A rectangular patch of wild flowers which has a length of 4.5 yards and a width of 3.2 yards.

Name: _____   Date: _____

Instructor: _____   Section: _____

Notebook 8.7
Understanding Circles

Define a circle.

What is the difference between a radius and a diameter?

    radius—distance from _____ to _____.

    diameter—distance _____.

Label each part of a circle.

What is the relationship between the diameter and the radius?

What are the formulas for diameter and radius?

EXAMPLE 1 Use the appropriate formula to find the radius of a pizza with a diameter of 10 inches.

EXAMPLE 2 Which situation involves diameter and which involves radius?
    A neighbor purchased a 14-foot circular trampoline for his children.

    In softball, the circle around the pitcher's mound is drawn 6 feet from where the pitcher stands.

What is circumference?

What is pi?
Ex.

131

What are the two approximations we can use for π?

CONCEPT CHECK:
A wheel makes one revolution on the ground. What part of the circle can be used to determine the distance traveled?

GUIDED EXAMPLES:
1. Does the situation described involve the diameter of the radius?
   A toy helicopter has moveable blades that are 5 inches long.

2. Does the situation described involve the diameter of the radius?
   A child's circular board game folds along the center with a measure of 15 inches.

3. Find the radius of a circular garden with a diameter of 3.4 feet.

4. Find the radius of a circular drum with a diameter of 25 cm.

Notebook 8.8
Finding Circumference

The _____ is the distance around the circle.
What is π?

The formula for finding the circumference of a circle is different depending on what is given.
What is the formula for finding the circumference of a circle,
… when you are given the radius?

…when you are given the diameter?

What are the units for circumference?

Exact answers for circumference are in terms of _____.
Approximate answers can be obtained using _____ or _____ for π.

EXAMPLE 1 Find the circumference of a circle with a radius of 8 inches. Find the exact answer in terms of π. Then find the approximate value using 3.14 for π.

EXAMPLE 2 Find the circumference of a circle with a diameter of 5 centimeters. Find the exact answer in terms of π. Then find the approximate value using 3.14 for π.

EXAMPLE 3 Find the circumference of a circle with a diameter of 21 yards. Use the approximate value of $\dfrac{22}{7}$ for π.

EXAMPLE 4 Find the circumference of a circle with a radius of 3.5 miles. Use the approximate value of 3.14 for π.

EXAMPLE 5 The earth's equator forms a circle. Estimate the number of miles a ship would have to travel if it went around the earth at the equator. Use 7900 miles as an approximation of the earth's diameter. Use 3.14 for π.

Understand the problem

Create a plan

Find the answer

Check the answer

CONCEPT CHECK:

A circle has a diameter of 10 feet. What is the first calculation that would be performed to find the circumference. If needed, use 3.14 for π.

GUIDED EXAMPLES:

Find the circumference of the following circles. Find the exact answer in terms of π. Then find the approximation of 3.14 for π. Round to the nearest tenth.

1. A circle with diameter of 3 feet.

2. A circle with a radius of 4 inches.

Find the circumference of each circle with the given diameter. Use $\dfrac{22}{7}$ for π.

3. 21 cm

4. 14 in.

Name: _____   Date: _____

Instructor: _____   Section: _____

Finding Area—Circles

OTHER NOTES

The area of a figure is the measure of the _____ inside the figure.

Area is measured in _____ of length.

What is the formula for finding the area of a circle with radius, *r*?

That is, to find the area, multiply _____ by _____.

What are the steps for finding the area of a circle?
1. Find the _____.
2.
3.

EXAMPLE 1 Find the area of a circle with a radius of 6 feet.
Use the approximate value of 3.14 for π.

EXAMPLE 2 Find the area of a circle with a diameter of 6 meters.
Use the approximate value of 3.14 for π.

EXAMPLE 3 Find the area of a circle with a diameter
of 7 feet. Use the approximate value of $\frac{22}{7}$ for π.

EXAMPLE 4 The diameter of a quarter is 2.5 cm. Find the area of a quarter, rounded to the nearest hundredth. Use the approximate value of 3.14 for π.

Understand the problem

Create a plan

Find the answer

Check the answer

CONCEPT

What is the formula for finding the area of a circle?

GUIDED EXAMPLES:

Find the area of the following circles.

1. A circle with a radius of 12 feet. Use 3.14 for π.

2. A circle with a diameter of 28 cm. Use $\dfrac{22}{7}$ for π.

3. A circle with a radius of 0.3 meters. Use 3.14 for π. Round to the nearest thousandth.

4. A circle with a diameter of 4 ft. Use 3.14 for π.

Notebook 9.1
Volume Definitions and Units

The distance around the figure is the _____.
Perimeter is measured in units of _____.
The measure of the surface inside the figure is the _____.
Area is measured in _____.

Complete the following table:

Measurements and Dimensions			
Measurement	Number of Dimensions	Dimensions	Label in Feet

Note: The exponent on the unit is equal to _____.

What is volume?
How do we measure volume?
Consider this: "How many _____ does it take to fill the object completely?"

Define rectangular box.

Define cube.

What is the volume of a cube that measures 1 inch on each side?
What is the volume of a cube that measures 1 centimeter on each side?

Which of these situations, if any would involve volume?
    Tiles on a floor
    Fencing around a garden
    Water inside a barrel

Which of these units, if any, could represent volume?
    65 miles
    6 cubic yards
    12 square inches

Write the steps for finding the volume of a box by counting cubes.
1.

2.

3.

EXAMPLE 1 Find the volume of a box with a length of 4 inches, a width of 3 inches, and a height of 4 inches. You will need to draw the picture.

EXAMPLE 2 Find the volume of a cube that measures 4 cm on each side. You will need to draw the picture.

EXAMPLE 3 Find the volume of a box with length 5 m, width 1 m, and height 3 m. Draw the picture.

CONCEPT CHECK:
Volume involves working in how many dimensions?

GUIDED EXAMPLES:
1. Find the volume of a rectangular box with length 2 ft, width 4 ft, and height 3 ft.

2. Find the volume of a cube with the given side measure 3 cm.

3. Find the volume of a rectangular box with length 5 in., width 2 in., and height 4 in.

Name: _____   Date: _____

Instructor: _____   Section: _____

Notebook 9.2
Finding Volume

The amount of space inside a figure is the _____.

Volume is measured in _____.

A _____ is a three-dimensional figure
   with 6 sides, and each side is a _____.

A _____ is a rectangular box in which every side is
   a _____.

We will learn formulas for finding volume of a rectangular box, a cube, a cylinder, and a sphere.

What is the formula for finding the volume of a rectangular box?
You will need to draw the box and label the sides.

EXAMPLE 1 Use the formula to find the volume of a rectangular box with a length of 12 inches, a width of 10 inches, and a height of 5 inches. You will need to draw and label the box.

What is the formula for finding the volume of a cube?
You will need to draw the cube and label the sides.

EXAMPLE 2 Use the formula to find the volume of a cube that measures 2.5 km on each side. You will need to draw and label the box. Round to the nearest hundredth.

What is the formula for finding the volume of a cylinder?
You will need to draw the cylinder and label the height and radius.

139

EXAMPLE 3 Use the formula and the approximate value of 3.14 for π to find the volume of a cylinder with a radius of 3 inches and a height of 5 inches. Draw your cylinder and label the radius and height.

What is the formula for finding the volume of a sphere?
You will need to draw the sphere and label the radius.

EXAMPLE 4 Use the formula and the approximate value of 3.14 for π to find the volume of a sphere with a radius of 6 mm. Draw your sphere and label the radius.

EXAMPLE 5 Find the volume of a soda can (in cubic centimeters) with a diameter of 6.6 cm and a height of 12.1 cm. Use the approximate value of 3.14 for π. Round to the nearest tenth.
Understand the problem

Create a plan

Find the answer

Check the answer

CONCEPT CHECK: The formula $V = \pi r^2 h$ is used to find the volume of what object?

GUIDED EXAMPLES:
1. Find the volume of a rectangular box with length of 4.5 ft, width of 3 ft and a height of 6 ft.

2. Find the volume of a cube with a side length of 7 inches.

3. Find the volume of a cylinder with a radius of 5 cm, and a height of 4 cm. Use 3.14 for π.

4. Find the volume of a sphere with a radius of 3 mm. Use 3.14 for π.

Notebook 9.3
Square Roots

What does it mean if a number is squared?

What is a perfect square?

List the first 15 perfect squares:

Is 81 a perfect square?   Why or why not?

Is 50 a perfect square?   Why or why not?

Is 64 a perfect square?   Why or why not?

Is 27 a perfect square?   Why or why not?

When you multiply two identical factors to result in another number,
the _____ of the number is one of those _____.

Finding a square root is the reverse operation of _____.

What is the symbol for finding the square root?
What is this symbol called?

Define square root.

EXAMPLE 1 Find the square root of 36.

EXAMPLE 2 Simplify $\sqrt{64}$.

EXAMPLE 3 Simplify $\sqrt{400}$.

EXAMPLE 4 The area of a square is 49 cm^2.
What is the length of each side?

EXAMPLE 5 Use a calculator to approximate $\sqrt{72}$.
Round to the nearest hundredth.

What is another way to approximate the square root of a number?

EXAMPLE 6 Approximate $\sqrt{75}$ by finding the two consecutive whole numbers between which the square root lies.

CONCEPT CHECK:

To approximate $\sqrt{34}$, between which two consecutive whole numbers would the square root be?

GUIDED EXAMPLES:

1. Identify whether the following is a perfect square.   39

2. Simplify the square root. $\sqrt{121}$

3. Use a calculator to approximate each square root. Round to the nearest hundredth. $\sqrt{200}$

4. Find the two consecutive whole numbers between which the square root lies. $\sqrt{51}$

142

Name: _____   Date: _____

Instructor: _____   Section: _____

Notebook 9.4
The Pythagorean Theorem

How many right angles are in a right triangle?

What are the legs of a right triangle?
Which side is the hypotenuse?
Label each side in the triangle.

State the Pythagorean Theorem.

What is the Pythagorean Theorem used for?

Write the steps for finding the missing side in a right triangle.
1. Identify the _____, the _____ and the _____.
To find the hypotenuse, use …
To find one of the legs, use …

2.

3.

EXAMPLE 1 Find the length of the hypotenuse of a right triangle with legs
that measure 3 cm and 4 cm.

EXAMPLE 2 One leg of a right triangle measures 9 inches and the hypotenuse
measures 15 inches. Find the length of the other leg.

EXAMPLE 3 Find the length of the missing side of the right triangle. Round
to the nearest tenth.

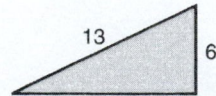

143

EXAMPLE 4 A standard computer monitor has a length of 20 inches and a width of 15 inches. What is the measure of the diagonal of the computer monitor?

Understand the problem

Create a plan

Find the answer

Check the answer

CONCEPT CHECK:
The Pythagorean Theorem is used to find the length of an unknown length of a side of what kind of triangle?

GUIDED EXAMPLES:
1. A right triangle has legs of lengths 5 feet and 12 feet. Find the length of the hypotenuse.

2. A right triangle has a leg of length 6 inches and a hypotenuse of length 10 inches. Find the length of the other leg.

3. A right triangle has a leg of length 5 m and a hypotenuse of length 7 m. Find the length of the other leg. Round to the nearest tenth.

4. Jacqueline's triangular flower garden has one side along the house and the other by the driveway, meeting the house at a right angle. The house side of the garden is 12 feet long, and the driveway side is 9 feet long. How much edging material will she need for the side that runs from the house to the driveway, which is the hypotenuse of the right triangle?

Notebook 9.5
Similar Figures

What does it mean if two figures are similar?
Draw similar figures.

True/False Similar figures must have the same shape
and the same size.

A _____ is a statement that two ratios are _____.
What does a proportion look like? _____ = _____.

How do you determine if two figures are similar?

For similar figures, the _____ of corresponding sides are
_____, their lengths have the same _____.

EXAMPLE 1 Identify the corresponding sides and set up proportions to
determine if the figures are similar. You will need to label the sides of the
triangles.

EXAMPLE 2 Identify the corresponding sides and set up proportions to
determine if the figures are similar.

All squares are _____.

EXAMPLE 3 Determine if the figures are similar.

145

EXAMPLES 4 & 5 Determine if the figures in each pair are similar. You will need to label the sides.

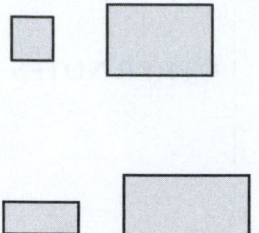

CONCEPT CHECK:

What does it mean when "the lengths of corresponding sides of similar figures are proportional"?

GUIDED EXAMPLES: Similar Figures

1. Determine if the figures are similar.

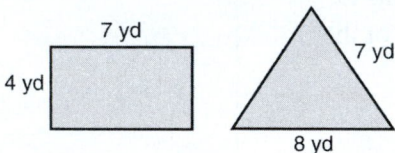

2. Identify corresponding sides in the similar triangles below.

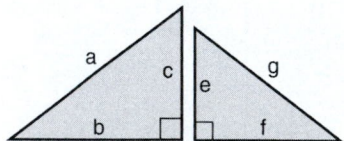

3. Determine if the figures are similar.

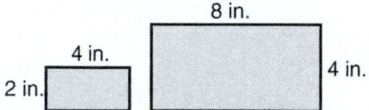

4. Determine if the figures are similar.

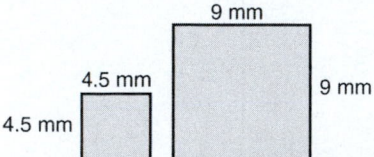

146

Name: _____   Date: _____

Instructor: _____   Section: _____

Notebook 9.6
Finding Missing Lengths

What does it mean if two figures are similar?

Corresponding sides of _____ figures are _____.

EXAMPLES 1 & 2 Write a proportion for the corresponding sides of the similar figures. You will need to label the sides.

Write the steps for finding the unknown side lengths in similar figures.
1.

2.

EXAMPLE 3 Find the unknown side length x in the similar figures. You will need to label the sides.

EXAMPLE 4 Find the unknown side length x in the similar figures.

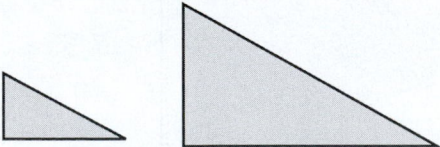

EXAMPLE 5 Find the unknown side length x in the similar figures. You will need to label the sides.

EXAMPLE 6 Kristen takes a 4-inch wide by 6-inch tall 4" x 6" photo of her family to a photography lab to be enlarged into a poster that will be 48 inches tall. Find the smaller dimension, or width, of the poster.

Understand the problem

Create a plan

Find the answer

Check the answer

CONCEPT CHECK:

Chef Aiko has two similar rectangular cutting boards. One has a length of 11 inches and a width of 9 inches. The other has a width of 6 inches. Set up the proportion to find the unknown length of the second cutting board, x.

GUIDED EXAMPLES:

1. Find the unknown side length in the similar figures

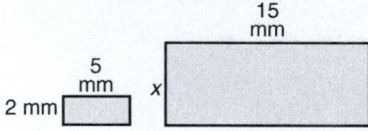

2. Find the unknown side length in the similar figures.

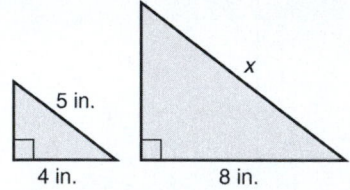

3. Find the unknown side length in the similar figures.

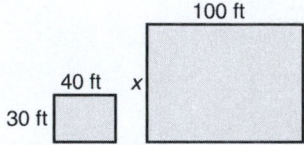

4. Marian took a 3" × 5" photo to a shop to be enlarged into a poster 40" tall. Find the smaller dimension, or width, of the poster.

148

Name: _____   Date: _____

Instructor: _____   Section: _____

Notebook 9.7
Congruent Triangles

What does it mean if two objects are congruent?

Note. Two triangles are _____ if the measures of corresponding _____ are equal and the _____ of corresponding sides are the _____.

State the rule for congruence by side-side-side (SSS).

EXAMPLE 1 Explain why $\triangle ABC$ and $\triangle EFD$ are congruent.

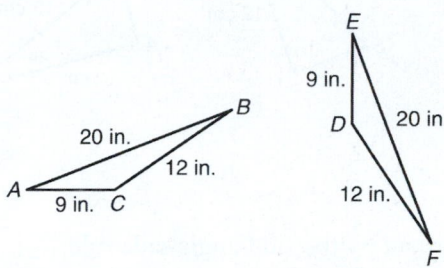

State the rule for congruence by side-angle-side (SAS).
Two triangles are congruent by SAS if the following conditions are met:
1.

2.

EXAMPLE 2 Explain why $\triangle STU$ and $\triangle CAB$ are congruent.

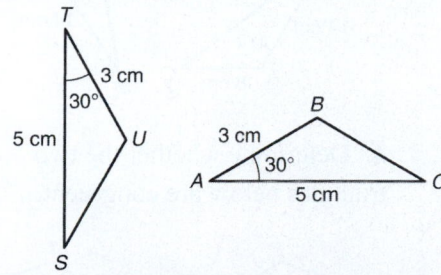

State the rule for congruence by angle-side-angle (ASA).
Two angles are congruent by ASA if the following conditions are met:
1.

2.

EXAMPLE 3 Explain why △*ABC* and △*STU* are congruent.

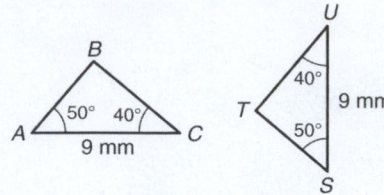

Do all combinations of congruent sides and angles show congruent?
Tell why or why not.

EXAMPLES 4 & 5 Determine whether each pair of triangles is congruent.

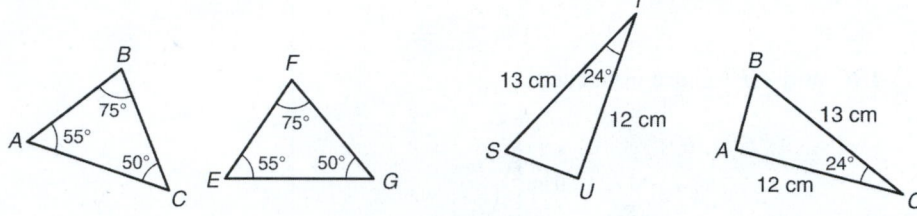

CONCEPT CHECK:
Describe two triangles that are congruent by the side-angle-side rule.

GUIDED EXAMPLES:

1. Explain why the two triangles below are congruent.

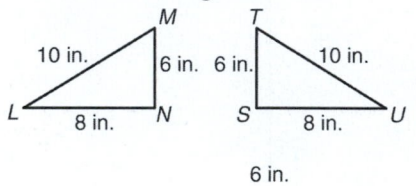

2. Explain why the two triangles below are congruent.

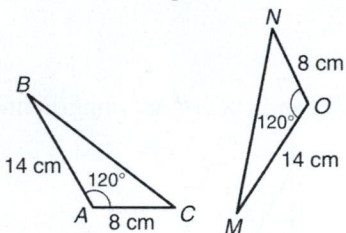

3. Explain why the two triangles below are congruent.

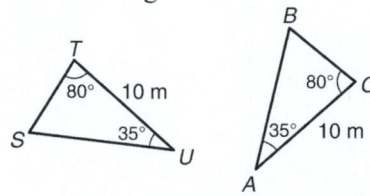

4. Determine whether the two triangles below are congruent.

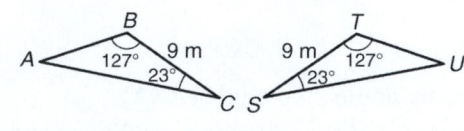

Name: _____   Date: _____

Instructor: _____   Section: _____

Notebook 9.8
Applications of Equations and Geometric Figures

What are the four basic steps for problem solving?

1.

2.

3.

4.

EXAMPLE 1 Find the unknown width of the rectangular solid.

Understand the problem.
Create a plan.
Find the answer.
Check the answer.

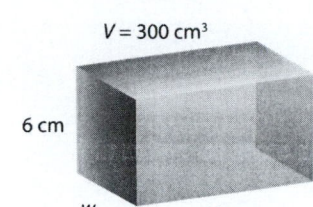

$V = 300 \text{ cm}^3$

6 cm

w 10 cm

EXAMPLE 2 Find the length of the guy wire supporting the telephone pole.

Understand the problem.
Create a plan.
Find the answer.
Check the answer.

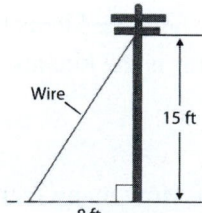

Wire

15 ft

8 ft

EXAMPLE 3 A manhole cover has a diameter of 3 ft. A strip of brass runs around the circumference of the cover. What is the length of this brass strip? Round to the nearest hundredth. (Use 3.14 for π.)

Understand the problem.
Create a plan.
Find the answer.
Check the answer.

151

EXAMPLE 4 A flagpole casts a shadow. At the same time, a small tree casts a shadow. Use the sketch to find the height of the flagpole.

Understand the problem.
Create a plan.
Find the answer.
Check the answer.

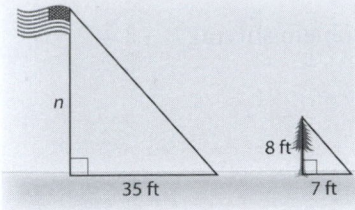

CONCEPT CHECK:
What formula is used to find the width of a rectangular solid?

GUIDED EXAMPLES:
1. Find the unknown length of the rectangular solid.

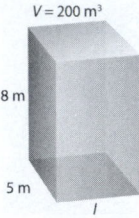

2. Barbara is flying her dragon kite on 32 yd of string. The kite is directly above the edge of a pond that is 30 yd from where the kite is tied to the ground. To the nearest tenth of a yard, how far is the kite above the pond?

3. Jimmy's truck has tires with a radius of 30 in. How many feet does his truck travel if the wheels make 9 revolutions? (Use 3.14 for π.)

4. A flagpole casts a shadow. At the same time, a small tree casts a shadow. Use the sketch to find the height n of the flagpole.

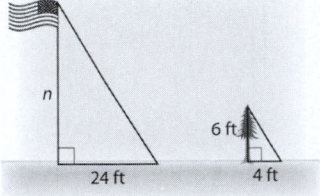

152

Notebook 10.1
Bar Graphs

A bar graph can:
    display _____
    show _____ over a period of _____
    show how often a particular data value _____
    compare _____ types of _____
    use _____ to distinguish between
        different _____ of data

EXAMPLE 1 Use the bar graph to answer the following questions.

1. How many total medals did the U.S. win in the 2008 Olympics?

2. Which type of medal did the U.S. win most often?

3. How many more silver medals were won than bronze medals?

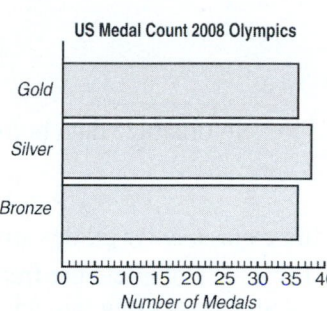

EXAMPLE 2 Use the bar graph to answer the following questions.

1. How many Country Music Association Awards did Loretta Lynn win?

2. In which category did she win the fewest awards?

3. Write the ratio of the number of Entertainer of   the Year awards won to the total number of   awards won.

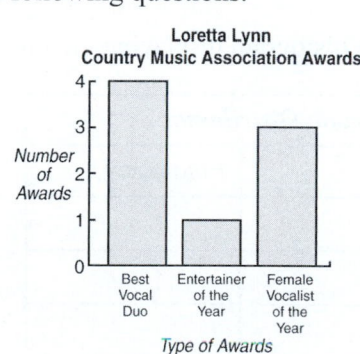

EXAMPLE 3 Use the double bar graph to answer the following questions.

1. In which game did the Phillies score the most runs?

2. In which game(s) did the Yankees score more runs than the Phillies?

3. In which game(s) did the largest difference between the numbers of runs scored by the two teams occur?

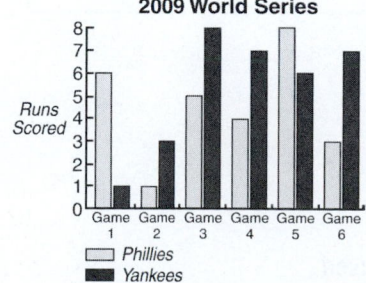

153

A histogram is a special type of _____.
In a histogram, the width of the bar represents an _____,
or _____, of numbers.
How does a bar graph differ from a histogram?

EXAMPLE 4 Use the histogram to answer the following questions.

a. What grade was earned most often?
   How many times?

b. How many students were in the class?

c. How many students earned an F for the
   course?

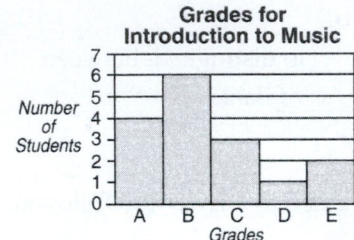

What kind of graph can be used to help organize data before creating a
histogram?

EXAMPLE 5 In an Intro to Music class, letter grades are distributed based on
the scale in the first column of Table 1. Complete the frequency distribution
table for an Intro to Music class with scores of 78, 69, 82, 95, 92, 80, 47, 89, 81,
and 99.
Then use the table to create a histogram displaying the same data.

TABLE 1 *Intro to Music Grade Distribution*

Grade Intervals	Tally	Frequency
90–100		
80–89		
70–79		
60–69		
Below 60		

CONCEPT CHECK:
Why should a bar graph be used?

Name: _____   Date: _____

Instructor: _____   Section: _____

GUIDED EXAMPLES:

1. The bar graph shows the numbers of active coal mines in 2007 for several states.

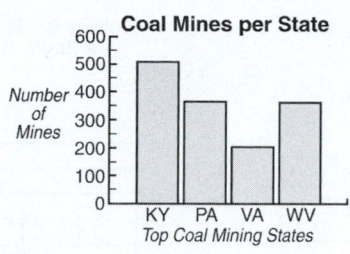

a. Which state had the most mines? How many mines did it have?

b. Which state had the fewest mines? How many mines did it have?

c. Which states had the same number of miens? How many mines did each have?

2. The bar graph to the right shows the numbers of medals won by U.S. men and women and women at the World Figure Skating Championships.

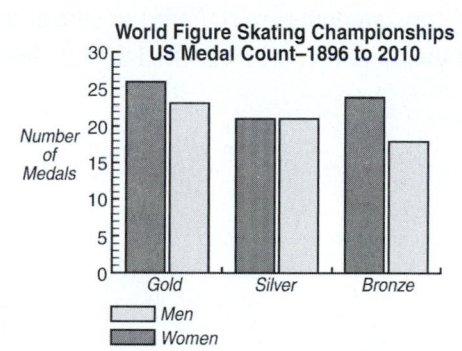

a. Who won more gold medals—men or women? How many more gold medals were won?

b. Which type of medal was won the same number of times by both men and women? How many times was it won?

c. Which type of medal was won by the U.S. men the fewest times? How many times was it won?

3. The following frequency distribution table and histogram show the number of college credit hours taken by a group of 20 seniors.

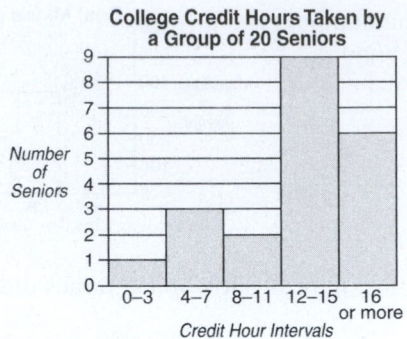

College Credit Hours Taken by a Group of 20 Seniors		
Credit Hour Intervals	Tally	Frequency
0–3	I	1
4–7	III	3
8–11	II	2
12–15	卌 IIII	9
16 or more	卌 I	6

a. How many more seniors were taking 12–15 credit hours than were taking 4–7 credit hours?

b. How many seniors are taking 16 or more credit hours?

c. At most colleges, full-time students are required to take at least 12 credit hours. How many of these seniors are classified as full-time students?

156

Name: _____   Date: _____

Instructor: _____   Section: _____

Notebook 10.2
Line Graphs

A line graph can:

    display _____

    show _____ over a period of _____

    show how often a particular data value _____

    compare _____ types of _____

    use data points that are connected with _____

OTHER NOTES

EXAMPLE 1 Use the line graph to answer the following questions.

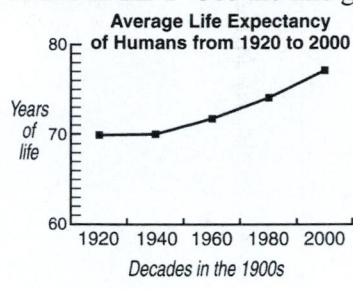

1. Has the average life expectancy of humans increased or decreased since 1920??

2. During which 20-year interval did the average life expectancy increase the most?

EXAMPLE 2 Use the line graph to answer the following questions.

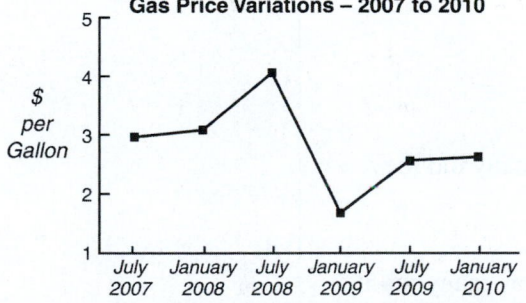

1. During which interval(s), or period(s) of time, did the largest increase/decrease in gas prices occur?

2. During the first half of 2009, did gas prices increase or decrease?

3. During which month and year does the graph show the highest gas price?

What kind of graph is used to show more than one set of data on a graph?

157

EXAMPLE 3 Use the double line graph to answer the following questions.

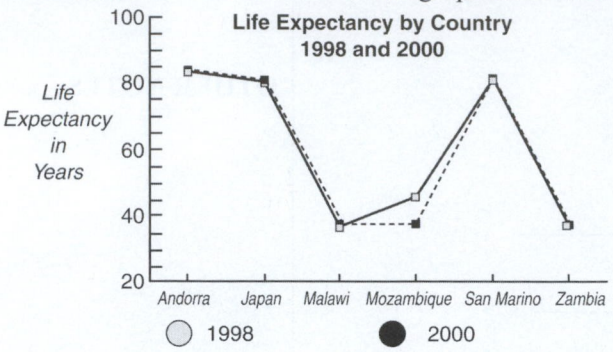

1. Which countries had the same life expectancy for 2000 and 1998?

2. What was the approximate life expectancy for Japan in 2000?

3. In which year was Mozambique's life expectancy lower, 2000 or 1998?
   How much lower was the life expectancy?

CONCEPT CHECK:

What is the difference between a bar graph and a line graph.

GUIDED EXAMPLES:

1.

US Winter Olympics Medal Count 1994–2010

a. In which year did the U.S. win the most medals? How many did they
   win?

b. In which year did the U.S. medal count decrease? By how many did it
   decrease?

c. In which year did the U.S. medal count increase the most over the
   previous Olympics? By how many?

d. In which years were the U.S. medal counts the same?

158

2.

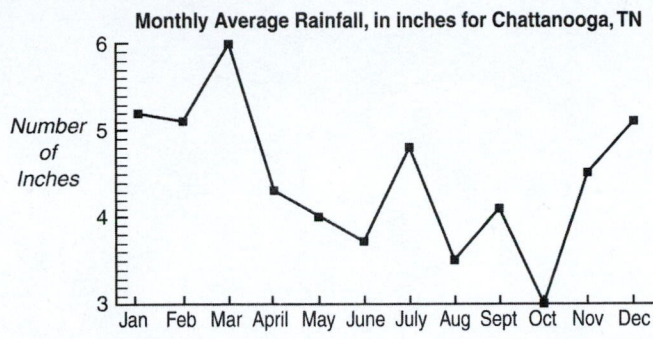

Monthly Average Rainfall, in inches for Chattanooga, TN

a. Which month had the highest average rainfall? What was the amount of rainfall?

b. Which month had the lowest average rainfall? What was the amount?

c. Between which two consecutive months did the greatest increase in rainfall occur? What was the difference in these two amounts?

3.

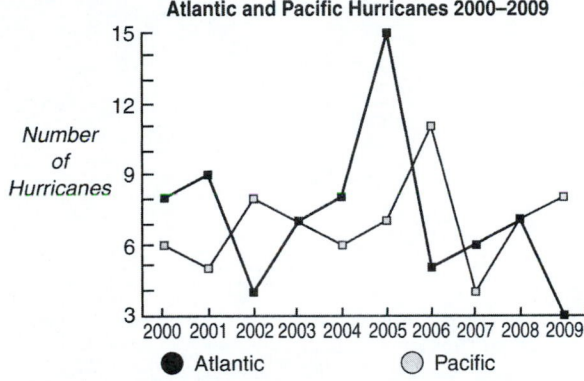

Atlantic and Pacific Hurricanes 2000–2009

● Atlantic   ○ Pacific

a. During which year did the most hurricanes occur in the Atlantic? How many hurricanes occurred in the Atlantic in that year?

b. During which year did the fewest hurricanes occur in the Pacific? How many hurricanes occurred in the Pacific in that year?

c. During which year(s) did more hurricanes occur in the Pacific than in the Atlantic?

d. During which year(s) did the Atlantic and the Pacific have the same number of hurricanes?

Name: _____   Date: _____

Instructor: _____   Section: _____

Notebook 10.3
Circle Graphs

A circle graph can:
   be used to show how _____
   be drawn using actual _____ or _____

Notes:
In circle graphs, the circle represents the _____ or _____ amount.
All sections must combine to give the _____ or _____.

EXAMPLE 1 Nikeshia received a scholarship at a local community college.
The circle graphs shows how she plans to use the money.

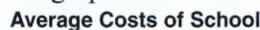

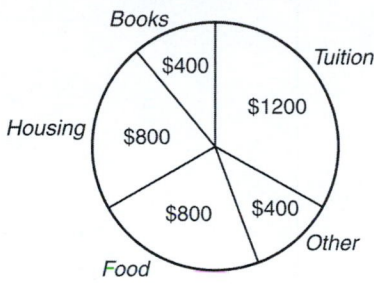

1. What does Nikeshia plan to spend the most money on?

2. What total amount does Nikeshia plan to spend on food and housing?

3. How much more does she plan to spend on tuition than on housing?

4. What is the total amount, in dollars, of Nikeshia's scholarship?

161

EXAMPLE 2 A Wildlife Management major at Cleveland State University needs 60 total credit hours to graduate with an Associate's Degree. The circle graph shows the number of credit hours needed in each category.

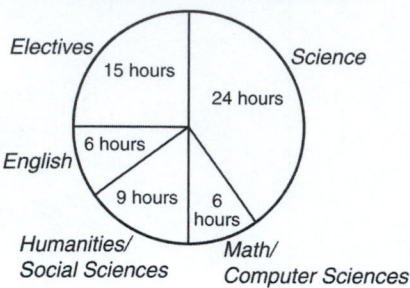

Required credit hours for
Wildlife Management Associate's Degree

1. How many credit hours of electives are required for the degree?

2. What is the ratio of credit hours in Math/Computer Sciences to credit hours in Science?

3. How many credit hours are needed in English and Humanities/Social Sciences?

EXAMPLES 3 The circle graph shows how a student's final grade is calculated according to one instructor's syllabus.

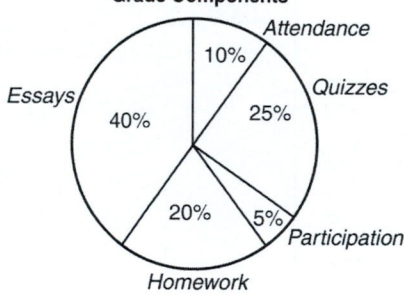

Breakdown of Writing Class
Grade Components

1. Which assignment type counts the most toward the student's final grade?

2. Assuming a student can earn a total of 2000 possible points in the class, how many points can be earned for attendance?

3. Assuming a student can earn a total of 1500 possible points, how many points can be earned for attendance?

CONCEPT CHECK:
True/ False   A circular graph can be used with more than one set of data.

162

Name: _____   Date: _____

Instructor: _____   Section: _____

GUIDED EXAMPLES:

1. The circle graph shows how 70 people rated an album on Amazon.com from 1 (lowest) to 5 (highest).

   a. What was the most popular rating?
   b. How many people rated the album as a 4?
   c. What is the ratio of the number of

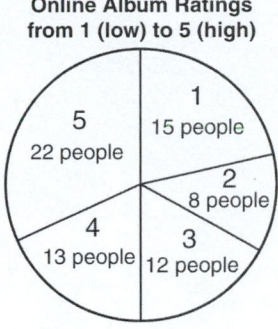

2-ratings to the number of 3-ratings?
   d. How many people rated the album as a 1 or a 2?

2. The circle graph shows how Marcie spent her day.

   a. How many hours did Marcie sleep?
   b. How did Marcie spend the greatest number of waking hours?
   c. How many hours did she spend on school,

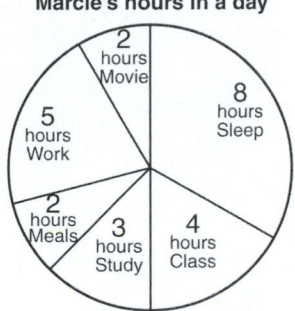

either in class or studying for class?
   d. How many more hours did she spend working than studying?

3. The circle graph shows U.S. mineral imports by country in 1993.

   a. What percent of minerals did the U.S. import from Mexico?
   b. What was the percent difference in the amount of mineral imports from Saudi

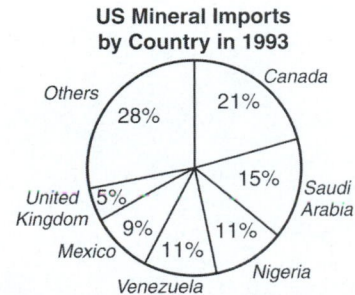

Arabia and the amount of mineral imports from the United Kingdom?
   c. From which two countries did the U.S. import the same percent of minerals?

163

Name: _____   Date: _____

Instructor: _____   Section: _____

Notebook 10.4
Mean

What is statistics?
_____ is information about a group or topic, often consisting
of a set of _____.

What are three measurements commonly used to study the behavior
of a set of numbers?
1.
2.
3.

These three measures are often referred to as _____
_____. Why??

The mean is the _____ of the data.
The median is the _____ of the data.
The mode is the number(s) occurring _____ in the data.

Define mean.
The mean is the _____ of the values in the list _____
the number of _____.

      Mean = _____

How do you calculate the mean? Write the steps.
1.

2.

EXAMPLE 1 Calculate the mean of the following list of numbers.

    3, 5, 7, 2, 11, 6, 8, 2, 10
  Sum =

  Number of values =

EXAMPLE 2 Calculate the mean of the following list of test scores.
  93, 86, 95, 98, 82

Sum =

Number of values =

EXAMPLE 3 Eli saves a portion of his allowance for 6 weeks. He saves $14 the first week, $12 the second week, $6 the third week, and $4, $10, and $2 for the remaining weeks. On average, how much does he save each week?
Understand the problem.

Create a plan.

Find the answer.

Check the answer.

CONCEPT CHECK:
Explain how to find the mean of 10, 15, and 20.

GUIDED EXAMPLES:
Calculate the mean of each list of numbers.
1. 7, 3, 11, 9, 14, 16, 20

2. 15, 21, 17, 22, 19, 25, 10, 22, 9, 16

3. Lynn went shopping and purchased items for $14.98, $7.60, $19.99, 11.95, and $8.53. Find the average cost of her purchases.

Notebook 10.5
Median

_____ is the study and use of data.

_____ is information about a group or topic, often consisting of a set of

_____ .

The _____ , or average, is one statistical measure we can use when working
with data.

The _____ , or middle number, is another.

What is the median?

What if there are two middle numbers?

Write the steps for finding the median of a list of numbers.
1.

2.

3.

Note: The numbers must be written in order from _____ to
_____ before finding the median.

EXAMPLE 1 Calculate the median of the following list of numbers.

   3, 5, 7, 2, 6, 8, 2, 10, 11
Ordered list:

Odd/even?

Middle number:

EXAMPLE 2 Calculate the mean of the following list of test scores.
   44, 32, 31, 56, 77, 65
Ordered list:

Odd/even?

Average of two middle numbers:

167

EXAMPLE 3 A poll of nine students shows that the number of Facebook friends they have are 395, 486, 166, 430, 172, 159, 723, 582, and 319. Find the median number of Facebook friends for these nine students.
Understand the problem.

Create a plan.

Find the answer.

CONCEPT CHECK:
What is the median?

GUIDED EXAMPLES:
Find the median of each list of numbers.
1. 2, 5, 3, 7, 9, 5, 1, 3, 6, 9, 5

2. 11, 15, 13, 19, 18, 16, 11, 14

3. Seven friends worked part time for a week to earn money for a camping trip. The amounts they earned were $66.00, $48.75, $86.90, $41.00, $46.50, $87.50, and $64.75. Find the median of the amounts earned.

Name: _____   Date: _____

Instructor: _____   Section: _____

Notebook 10.6
Mode

Statistics is the study and use of _____.
Data is information about a _____ or topic, often consisting of a set of
_____.

What is the mode?

How many modes can there be?

Notes:
Is it necessary to write the numbers in order before finding the mode?
Is a number that occurs more than once always a mode?

EXAMPLE 1 Find the mode(s) of the following list of numbers.

 3, 5, 7, 2, 11, 6, 8, 2, 10
Ordered list:

 Mode(s):

EXAMPLE 2 Find the mode(s) of the following list of numbers.
 44, 32, 31, 56, 77, 65 65, 44, 65
Ordered list:

Number(s) occurring more often:

Mode(s):

EXAMPLE 3 Find the mode(s) of the following list of numbers.
 325, 612, 367, 692, 451
Ordered list:

Number(s) occurring more often:

Mode(s):

EXAMPLE 4 Find the mode(s) of the following list of numbers.

    2.1, 5.3, 6.2, 6.2, 7.5, 8.1, 8.1, 9.0, 11.4

Ordered list:

Number(s) occurring more often:

Number(s) occurring most often:

EXAMPLE 5 According to the National Weather Service, the high temperatures (in degrees Fahrenheit) for the first two weeks of June 2010 were recorded as 105, 106, 109, 112, 116, 119, 119, 115, 111, 107, 106, 103, 104, and 110. Find the mode(s) of the high temperatures for the time period.

Understand the problem.

Create a plan.

Find the answer.

Check the answer.

CONCEPT CHECK:
How many modes can there be for a given list of numbers?

GUIDED EXAMPLES:
Calculate the mode of each list of numbers.
1. 2, 5, 3, 7, 9, 5, 1, 3, 6, 9, 5

2. 25, 18, 30, 28, 42, 33, 51, 12

3. Over the past ten years, the numbers of hurricanes in the Atlantic Ocean were: 8, 9, 4, 7, 8, 15, 5, 6, 7, and 3. Find the mode(s) of the numbers of hurricanes during this 10 year period.

Notebook 10.7
Introduction to Probability

What is the purpose of a tree diagram?
Draw a tree diagram for flipping a coin.

EXAMPLE 1 Draw a tree diagram showing the possible outcomes of rolling a single die.

What is probability?

What does a probability of 1 mean?
What does a probability of 0 mean?

How do you calculate probability?

EXAMPLES 2–4 Find the probability of the following events.
Tossing a coin and having it land on tails

Rolling a 4 on a single die

Rolling a 1 or a 3 on a single die

EXAMPLES 5–9 A bag contains 6 red marbles, 7 blue marbles, 2 black marbles, and 5 green marbles.
Find the probability of picking a red marble out of the bag.

Find the probability of picking a black marble out of the bag.

Find the probability of picking a blue or black marble.

Find the probability of picking a marble of any color.

Find the probability of picking a yellow marble.

EXAMPLES 10–12 A board game uses a colored spinner to determine where a player will move on his or her next turn.

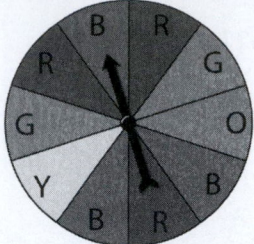

Find the probability of the spinner landing on red.

Find the probability of the spinner landing on green.

Find the probability of the spinner landing on a primary color.
(red, yellow, or blue).

CONCEPT CHECK:
True/False A tree diagram showing the outcomes of a 6-sided die will have 7 branches.

GUIDED EXAMPLES:

1. Draw a tree diagram showing the number of outcomes of throwing an octahedral (8-sided) die with each side representing a number 1 through 8.

2. Find the probability of the following event.
   Tossing a coin and getting heads

A bag contains 1 red marble, 5 green marbles, 2 blue marbles, and 2 black marbles. Find the probability of the following events.

3. Picking a blue marble

4. Picking a blue or green marble

Name: _____   Date: _____

Instructor: _____   Section: _____

Notebook 11.1
Introduction to Real Numbers

What are whole numbers?

What are integers?

_____ are numbers that can be written as a fraction of two integers. Give examples.

      Write a repeating decimal:

      Write a terminating decimal:

What are numbers that cannot be expressed as a fraction of two integers? Give examples.

True/False: All integers are rational numbers.

Rational numbers in decimal form are either _____

or _____.

The decimal form of an _____ is non-terminating and non-repeating. Give an example.

Define real numbers.

EXAMPLE 1 Classify each number as an integer, a rational number, an irrational number, and/or a real number.

Number	Integer	Rational Number	Irrational Number	Real Number
8				
$-\frac{1}{5}$				
-1.65				
$\sqrt{3}$				
0.444…				
$\sqrt{9}$				

Any _____ can be plotted on a number line.

173

The real number line includes _____,
_____, and _____.

Label the three parts of the real number line:

_____

EXAMPLES 2–4 Plot the following real numbers on a number line.

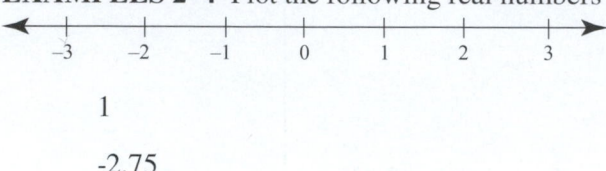

1

-2.75

½

Give some practical examples of real numbers.

EXAMPLES 5–7 Use a real number to represent each real life situation.
A temperature of 131.2° below zero is recorded in Antarctica.

The height of a mountain is 22,645 feet above sea level.

A golfer scored 5 under par in a recent tournament.

CONCEPT CHECK:
Give an example of a real-life negative number.

GUIDED EXAMPLES:
1. Classify the number as an integer, a rational number, an irrational number, and/or a real number. −6

2. Plot on a number line.
   $-\dfrac{2}{3}$

3. Plot on a number line.
   −0.75
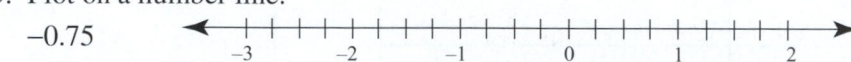

4. Use a real number to represent the real life situation.
   An oil drilling platform extends 325 feet below sea level.

174

Name: _____   Date: _____

Instructor: _____   Section: _____

Notebook 11.2
Graphing Rational Numbers Using a Number Line

Every real number can be _____ on a number line.

EXAMPLES 1 & 2 Plot the following real numbers on a number line.

2

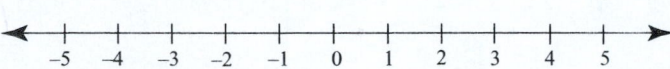

−1.5

An _____ is a statement that shows the relationship between any two real numbers that are not equal.

Which symbols are used to represent inequalities?
_____ is used to represent "is less than."

_____ is used to represent "is greater than."

EXAMPLES 3 & 4 Write the following statements using inequality symbols.

2 is less than 7

8 is greater than 5

Using a number line, one number is _____ another number if it is to the _____ of that number. Show this on a number line.

One number is _____ another number if it is to the _____ of that number on a number line. Show this on a number line.

True/False $3 > -1$ and $-1 < 3$ express the same meaning.

175

EXAMPLES 5–7 Plot the given numbers on a number line, and then replace the question mark with the appropriate symbol, < or >.

−2.5 ? 0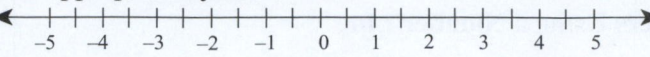

$4 \; ? \; \dfrac{3}{4}$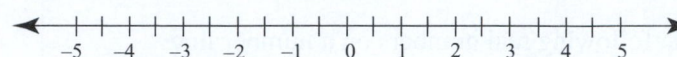

−1 ? −5

CONCEPT CHECK:

Correct or incorrect   0 > −1 can be read as "0 is less than −1"?

GUIDED EXAMPLES:

Plot each number on a number line.

1. −3.75

2. $\dfrac{3}{2}$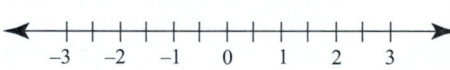

Write statements using inequality symbols.

3. −7 is less than 0

4. 3.5 is greater than 3

Name: _____   Date: _____

Instructor: _____   Section: _____

Notebook 11.3
Translating Phrases into Algebraic Inequalities

What is a variable?

To translate an everyday situation into an algebraic statement, each
_____ must be represented with a _____.

What is the symbol used for "is less than"?
What is the symbol used for "is greater than"?

What are the steps for translating words to symbols?
1. Determine …

2. Replace …

3. Replace …

EXAMPLE 1 Translate the phrase into an algebraic inequality.
A police office claimed that a car was traveling at a speed more than 85 miles per hour. (Use the variable s for speed.)

EXAMPLE 2 Translate the phrase into an algebraic inequality.
The bank loan officer said that the total consumer debt incurred by Hector and Erica must be less than $10,000 if they want to qualify for a mortgage to buy their first home. (Use the variable d for debt.)

What is the difference between the symbols $<$ and $>$ and the symbols $\leq$ and $\geq$?

What phrase is represented by the symbol $\leq$?
What phrase is represented by $\geq$?

177

Key words to determine which symbols are most appropriate:

Less than or equal to

Greater than or equal to

EXAMPLE 3 Translate the phrases into an algebraic inequality.
The owner of a trucking company said that the payload of a truck must be no more than 4500 pounds. (Use the variable p for payload.)

EXAMPLE 4 Translate the phrases into an algebraic inequality.
Carlos must be at least 16 years old in order to get his driving license. (Use the variable a for age.)

CONCEPT CHECK:
What symbol does the phrase "at most" or "no more than" translate to?

GUIDED EXAMPLES:
Translate the phrase into an algebraic inequality.
1. The height of a vehicle must be less than 7.5 feet in order to enter the parking garage. Use the variable h for the height.

2. According to the fire code, the room can hold at most 80 people. Use the variable n for the number of people.

3. A customer must be at least $4\frac{1}{2}$ feet tall in order to ride the roller coaster. Use the variable h for the customer's height.

4. During the heat wave, the high temperature was greater than 105° F. Use the variable T for the temperature.

Name: _____   Date: _____

Instructor: _____   Section: _____

Notebook 11.4
Finding the Absolute Value of a Real Number

Define absolute value.

EXAMPLES 1 & 2 Use a number line to find the absolute value of the following numbers.

3

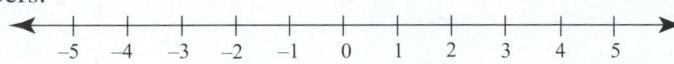

−1.5

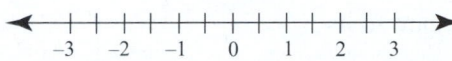

What is the symbol for absolute value?

How do you write the symbol form for absolute value of 5?

Write the absolute value of −2.87.

EXAMPLE 3 Write the expression for the absolute value of $-\dfrac{3}{8}$.

EXAMPLES 4–6 Find the absolute value of the following numbers.
|−3.68|

|3.68|

|0|

Note: Distance is always a _____, regardless of direction.
This means that the absolute value of any number will either be a
_____ _____ or _____.

EXAMPLES 7–9 Find the absolute value of the following numbers.
$|-8.333...|$

$$\left|5\frac{9}{10}\right|$$

$$\left|-\frac{2}{7}\right|$$

Note: The absolute value of a number can be thought of as the
_____ of the number, without regard to its
_____.

CONCEPT CHECK:
What is absolute value?

GUIDED EXAMPLES:
1. Find the absolute vale of −0.555.

2. Write the expression for the absolute value and then simplify. $\dfrac{4}{13}$

Simplify each absolute value.

3. $\left|-\dfrac{3}{10}\right|$

4. $|6.15|$

180

Name: _____   Date: _____

Instructor: _____   Section: _____

Notebook 12.1
Adding Real Numbers with the Same Sign

Banking Situation 1
You make a deposit of $20 on one day and a deposit of $15 the next day.
Represent on a number line the total value of your deposits.

Banking Situation 2
You write a check for $25 to pay your cell phone bell and two days later
you write a check for $5 to donate to a charity. Represent on a number line
the total value of the withdrawals.

Using the chip method, how do you represent a positive number?

Represent 3 using chips.
Represent –4 using chips.

EXAMPLES 1 & 2 Use chips to add the following.
$4 + 2$

$-2 + (-3)$

How do you add two numbers with the same sign?

EXAMPLE 3 Add. $8 + 4$
Add the numerical parts.

Give the answer the sign of the numbers being added.

Note: A number written without a sign is assumed to be _____.

181

EXAMPLE 4 Add. $-3 + (-5)$

EXAMPLE 5 Add. $-\dfrac{2}{3} + \left(-\dfrac{1}{7}\right)$

EXAMPLE 6 Add. $-8.1 + (-2.75) + (-5.03)$
Add from left to right.

CONCEPT CHECK:
When adding -0.25 and -0.66, what should be done first?

GUIDED EXAMPLES:
1. Add. $-3 + (-5)$

2. Add. $-5.18 + (-3.7)$

3. Add. $-\dfrac{3}{8} + \left(-\dfrac{3}{2}\right)$

4. Add. $-2 + (-7) + (-6)$

Notebook 12.2
Adding Real Numbers with Different Signs

Banking Situation 1
You make a deposit of $30 on one day, and on the next day you write a check for $25. Represent on a number line how the overall amount in your account changed after the deposit and withdrawal are processed.

Banking Situation 2
You make a deposit of $10 on one day, and on the next day you write a check for $40. Represent on a number line the overall change in the amount.

Using the chip method, a positive number is represented by _____
and a negative number is _____.

Represent 5 using chips.
Represent −3 using chips.
NOTICE: _____ + _____ = 0

EXAMPLES 1 & 2 Use chips to add the following.
6 + (−4)

−7 + 3

How do you add two numbers with different signs?

EXAMPLE 3 Add. −6 + 9
Subtract the numerical parts.

Give the answer the same sign
as the larger numerical part.

EXAMPLE 4 Add. $\dfrac{3}{8}+\left(-\dfrac{5}{7}\right)$

EXAMPLE 5 Add. $3.7+(-10.5)$

EXAMPLE 6 Add. $-\dfrac{3}{4}+\dfrac{7}{9}$

CONCEPT CHECK:

Determine whether the final answer to the addition $(-8)+7$ is negative or positive and indicate why.

GUIDED EXAMPLES:

Add the following.

1. $8+(-5)$

2. $-11+7$

3. $2+(-8)$

4. $-\dfrac{3}{11}+\dfrac{8}{11}$

Name: _____   Date: _____

Instructor: _____   Section: _____

Notebook 12.3
Finding the Opposite of a Real Number

Write the opposite of the following everyday situations:
A gain of 5 yards in a football game…

A check written for $28.50 on a checking account…

EXAMPLE 1 Find the opposite of the situation.
A temperature increase of 8.5°F
Opposite numbers are the _____ distance from zero but
in _____.

The opposite of 3 is _____.
The opposite of –3 is _____.

Represent the opposite numbers on the number line.
−2.16 and 2.16

$$-3 \quad -2 \quad -1 \quad 0 \quad 1 \quad 2 \quad 3$$

−3 and 3

$$-3 \quad -2 \quad -1 \quad 0 \quad 1 \quad 2 \quad 3$$

EXAMPLE 2 Find the opposite of the number using a number line. 8
Locate the number on a number line.

$$-10 \ -9 \ -8 \ -7 \ -6 \ -5 \ -4 \ -3 \ -2 \ -1 \ 0 \ 1 \ 2 \ 3 \ 4 \ 5 \ 6 \ 7 \ 8 \ 9 \ 10$$

Find the number that is the same distance from zero, but in the opposite side.

Note: The _____ of a number is the _____ of that number.
The symbol for "the opposite of" a number is the _____ (or
_____) sign, "_____".

If a is a real number, then the opposite of a, denoted _____ equals

_____.

The opposite of _____ is _____.

What is the step for finding the opposite of a number?

EXAMPLES 3–5 Find the opposite of the number.
7

−3.5

$-\dfrac{5}{7}$

185

The _____ of a number is its distance from zero on a number line. The symbol is _____.

EXAMPLES 6 & 7 Find the opposite of the following absolute values.

$|5|$

$|-6.78|$

Why do pairs of opposite numbers have the same absolute value?

Any number plus is opposite is equal to _____.

What are additive inverses?

EXAMPLE 8 Find the additive inverse, or opposite, of the number. Then add the additive inverse to the number.

-4

CONCEPT CHECK:
What is the correct translation of the statement "The opposite of the absolute value of -1 is -1"?

GUIDED EXAMPLES:
1. Find the opposite of -6 using a number line.

Find the opposite of each of the following:

2. $-\dfrac{2}{3}$

3. $|-2|$

4. Find the additive inverse, or opposite, of each number. Then add the additive inverse to the number. $-\dfrac{5}{8}$

186

Name: _____   Date: _____

Instructor: _____   Section: _____

Notebook 12.4
Subtracting Real Numbers

Let's review addition
Write the rule for adding two numbers with the same sign.

Write the rule for adding two numbers with different signs.

Write the rule for subtracting real numbers.

We sometime refer to this as _____,_____,_____.

What are the steps for subtracting real numbers?
1.
2.
3.
4.

EXAMPLE 1 Subtract. $10 - 40$
Leave the first number alone.
Change the minus sign to a plus sign.
Change the sign of the number being subtracted.
Add the two numbers using the rules of addition.

EXAMPLE 2 Subtract. $8 - (-3)$

EXAMPLE 3 Subtract. $-5 - (-9)$.

EXAMPLE 4 Subtract. $-6.18 - 2.34$

187

EXAMPLE 5 The temperature in a city is –8°F, and the temperature of a neighboring town is –15°F. Find the difference in temperature between the two cities.

CONCEPT CHECK:
Harry subtracted 7 – (–3) and got an answer of 4. This is incorrect. What was his error?

GUIDED EXAMPLES:
Subtract. Use "Add the Opposite."
1. 20 – 30

2. –2 – 9

Subtract. Use "Leave, Change, Change."
3. $\dfrac{13}{10} - \left(-\dfrac{7}{10}\right)$

4. –7 – (–2)

Notebook 12.5
Addition Properties of Real Numbers

What does the Commutative Property of Addition say?
_____ of the numbers being added does not change the sum.

Give examples:

What does the Addition Property of Zero say?
_____ to a number does not change the number.

Give examples:

What does the Associative Property of Addition say?
_____ when adding numbers does not change the sum.

Give examples:

Two numbers are _____, or _____, if they add to equal zero.

What does the Additive Inverse Property of Addition say?
The sum of a number and its additive inverse is equal to _____.
In symbols, for any real number a, $a + (-a) =$ ____ and $-a + a =$ _____.

Give examples:

EXAMPLES 1–5 Determine which property of addition is shown by each equation. The properties are the Commutative Property of Addition, the Associative Property of Addition, the Addition Property of Zero, and the Additive Inverse Property of Addition.

$0 + (-7) = -7$ _____

$(-2 + 3) + 5 = 5 + (-2 + 3)$ _____

$10 + (-10) = 0$ _____

$(-2 + 6) + 0.3 = -2 + (6 + 0.3)$ _____

$-\dfrac{2}{3} + \dfrac{1}{3} = \dfrac{1}{3} + \left(-\dfrac{2}{3}\right)$ _____

CONCEPT CHECK:

Which property is illustrated by $(-2) + 2 = 0$?

GUIDED EXAMPLES:

Determine which property of addition is shown by each equation. The properties are Commutative Property of Addition, the Associative Property of Addition, the Addition Property of Zero, and the Additive Inverse Property of Addition.

1. $-1 + (-9 + 5) = (-1 + (-9)) + 5$

2. $-5 + 6 = 6 + (-5)$

3. $7 + (-7) = 0$

4. $-\dfrac{1}{5} + \dfrac{1}{5} = 0$

190

Notebook 12.6
Applications of Adding and Subtracting Real Numbers

Let's review addition
Write the rule for adding two numbers with the same sign.

Write the rule for adding two numbers with different signs.

EXAMPLES 1–4 Add the following.

$5 + 9$ $-6 + 7$

$-2 + (-4)$ $1 + (-8)$

Write the rule for subtracting real numbers.
1.
2.
3.
4.

We sometime refer to this as _____,_____,_____.

EXAMPLES 5–8 Subtract the following.

$4 - 9$ $-2 - (-6)$

$-3 - 5$ $7 - (-8)$

EXAMPLE 9 Emmanuel will be driving several days to get home for a reunion. At one point on the first day of his trip, his elevation was 750 feet below sea level. The next day his elevation dropped another 450 feet. What was Emmanuel's elevation at the end of the second day?.
Understand the problem.
Create a plan.
Find the answer.
Check the answer.

EXAMPLE 10 At the end of June, the rainfall for the year, so far, was 6.5 inches below normal. During the second half of the year, it rained 11.25 inches above normal. What amount of rainfall fell above or below normal for the entire year?

Understand the problem.

Create a plan.

Find the answer.

Check the answer.

EXAMPLE 11 The Wildlife Club has an account balance of $61.46. They need to pay expenses of $99.98. How much money in membership dues do they need to collect first to avoid a negative account balance (that is, to have a zero balance after expenses are paid)?

Understand the problem.

Create a plan.

Find the answer.

Check the answer.

EXAMPLE 12 While playing golf, Don recorded the following scores on his first six holes: $+1, -2, -1, 0, +3,$ and $+1$. What was his total score after these first six holes?

Understand the problem.

Create a plan.

Find the answer.

Check the answer.

CONCEPT CHECK: A mountain is 14,050 feet above sea level and a valley is 143 feet below sea level. Write the expression that will calculate the difference in elevation between the mountain's peak and the bottom of the valley.

GUIDED EXAMPLES:

1. At 12:30 a.m., a submarine was 614 meters below the surface. Over the next three hours, the submarine continued to descend 127 meters deeper. How deep is the submarine at the end of the three hours?

2. In a certain city, the average temperature for January is $-9.8°$ Fahrenheit. The average temperature for February is $-2.6°$ Fahrenheit. What is the increase in the average temperature from January to February?

192

Notebook 13.1
Multiplying Real Numbers

Complete the sign chart for multiplying signed numbers:

+	+
+	−
−	+
−	−

This chart applies for both _____ and _____.

Complete the chart for multiplying with different signs:

3	·	2	=
3	·	1	=
3	·	0	=
3	·	−1	=
3	·	−2	=

Write the steps for multiplying two numbers with different signs.
To multiply two numbers with different signs,
1.

2.

EXAMPLE 1 Multiply. (7)(−4)
Multiply the numerical parts:
Determine the sign:

EXAMPLE 2 Multiply. $\left(-\dfrac{3}{4}\right)(5)$

Multiply the numerical parts:
Determine the sign:

Complete the chart for multiplying numbers with same sign:

−4	·	3	=
−4	·	2	=
−4	·	1	=
−4	·	0	=
−4	·	−1	=

Write the steps for multiplying two numbers with the same sign.
To multiply two numbers with the same sign,
1.

2.

193

EXAMPLE 3 Multiply. $(-12)(-9)$
Multiply the numerical parts:
Determine the sign:

EXAMPLE 4 Multiply $(-3.5)(-2)$
Multiply the numerical parts:
Determine the sign:

EXAMPLES 5–8 Multiply.
$(-3)(-4)$

$(-3)(-4)(-5)$

$(-3)(-4)(-5)(6)$

$(3)(-4)(-5)(6)$

Notes: When multiplying an even number of negative factors, the product is

_____.

When multiplying an odd number of negative factors, the product is

_____.

CONCEPT CHECK:
What will be the sign of the product below?
$(+)(-)(-)(+)(+)(-)$

GUIDED EXAMPLES:
Multiply.
1. $(14)(-8)$

2. $\left(-2\dfrac{2}{3}\right)(8)$

3. $(-3)(2)(5)(-4)$

4. $(-9)(4)(2)$

Name: _____   Date: _____

Instructor: _____   Section: _____

Notebook 13.2
Finding the Reciprocal of a Real Number

What are reciprocals?

$\dfrac{b}{a}$ is the reciprocal of _____.

The reciprocal is also called the _____.

How do you find the reciprocal of a number?

Note: The reciprocal will have the _____ as
the _____.

Ex.   What is the reciprocal of $\dfrac{3}{4}$?

        What is the reciprocal of $\dfrac{-13}{2}$?

A negative fraction can be written in _____ different
but _____ ways:

Note: How is a negative fraction typically written?

Examples of Negative Fractions

$\dfrac{-4}{2}$

$\dfrac{4}{-2}$

$-\dfrac{4}{2}$

EXAMPLE 1 Find the reciprocal. $-\dfrac{5}{7}$

Write the steps for finding the reciprocal of an integer.
1.

2.

What is the reciprocal of 0?
Why?

EXAMPLE 2 Find the reciprocal. -3

195

Write the steps for finding the reciprocal of a decimal.
To find the reciprocal of a decimal number,
1.

2.

3.

EXAMPLE 3 Find the reciprocal.   0.5

Write the steps for finding the reciprocal of a mixed number.
1.

2.

EXAMPLE 4 Find the reciprocal. $-4\dfrac{3}{5}$

CONCEPT CHECK:
True/False   To find the reciprocal of a fraction, invert (or flip) the fraction.

GUIDED EXAMPLES:
Find the reciprocal.

1. $\dfrac{9}{4}$

2. $-\dfrac{8}{11}$

3. -0.25

4. $-3\dfrac{3}{4}$

Name: _____   Date: _____

Instructor: _____   Section: _____

Notebook 13.3
Dividing Real Numbers

What is division?

What is the answer to a division problem?

The number being divided is the _____.
The number you are dividing by is the _____.

What are the two ways we can write division? Use the terms divisor, dividend, and quotient.

Write the procedure for dividing signed numbers?

How do you determine the sign of the answer (quotient)?

EXAMPLES 1–3 Divide.
$-21 \div (-7)$

$3.2 \div 0$

$\dfrac{-45}{25}$

EXAMPLE 4 Four friends decide to start a business together. They share a startup loan of $120,000. If they split the amount of the loan equally between them, how much does each friend owe?

Recall dividing fractions.
What is the rule for dividing by a fraction?

How do you divide by a mixed number?

197

EXAMPLES 5 & 6 Divide.

$$32 \div \left(-\frac{8}{3}\right)$$

$$-\frac{11}{6} \div 2\frac{4}{9}$$

CONCEPT CHECK:

Which of the following expressions has a negative quotient?

a. $\dfrac{0}{-12}$ 　　　 b. $\dfrac{-48}{-6}$ 　　　 c. $\dfrac{16}{-8}$ 　　 d. $\dfrac{35}{7}$

GUIDED EXAMPLES:

Divide.

1. $-90 \div (-18)$

2. $42 \div (-0.7)$

3. $\left(-\dfrac{4}{5}\right) \div \left(-\dfrac{7}{3}\right)$

4. $\left(-9\dfrac{1}{3}\right) \div \left(-3\dfrac{1}{2}\right)$

Name: _____   Date: _____

Instructor: _____   Section: _____

Notebook 13.4
Exponents and the Order of Operations

Why do we use exponents?
What is the base?
What is the exponent?

Write the exponential form of $2 \cdot 2 \cdot 2 \cdot 2$.

EXAMPLE 1 Write in exponential form.

$$\left(-\frac{5}{7}\right)\left(-\frac{5}{7}\right)\left(-\frac{5}{7}\right)\left(-\frac{5}{7}\right)$$

Caution! When the base is _____, be careful in determining the sign
of the _____-.

EXAMPLES 2 & 3 Evaluate.
$(-4)^2$

$(-4)^3$

Sign Rule for Exponents
     A _____ base raised to an even power is _____.
     A _____ base raised to an odd power is _____.

Caution! Be careful with exponents and negative signs.
     What does $(-2)^4$ mean?

     What does -2^4 mean?

     Do you see the difference???

EXAMPLE 4 Evaluate.

$$\left(-\frac{2}{5}\right)^3$$

Any non-zero number raised to the zero power is equal to _____.
Any number raised to the power of 1 is equal to _____.

EXAMPLES 5 & 6 Evaluate.
$(-94)^0$

$(-23)^1$

EXAMPLES 7 & 8 Evaluate.

$(-5)^2$

$-\left(-\dfrac{1}{5}\right)^3$

Write the steps for Order of Operations.
1. P:
2. E:
3. MD:
4. AS:

EXAMPLES 9 & 10 Evaluate.

$\dfrac{4^3 + 2(-5)}{2^3}$

$(-4)^2 - 2(5)^2$

EXAMPLE 11 Evaluate.

$2^6 \div 2^3 - [(3)^2 + 5 - (27)^0]^2$

CONCEPT CHECK:

For the expression -5^4, identify the base.

GUIDED EXAMPLES:

Evaluate.

1. $-\left(\dfrac{18}{73}\right)^1$

2. $(-0.6)^4$

3. $\dfrac{-3^3 + 2^2(7)}{-(-6)^2}$

200

Notebook 13.5
The Distributive Property

See egg carton example. Write down any notes you need.

OTHER NOTES

What does the Distributive Property say?

Examples
$-7(3 + 5)$

$2(5 - 7)$

EXAMPLE 1 Multiply. $9 \cdot 103$

Rewrite as an addition problem.
Use Distributive Property to multiply.
Simplify.

When is the distributive property most useful?

The Distributive Property can be used to _____.

EXAMPLE 2 Multiply. $2(x + 3)$

Rewrite using Distributive Property.
Answer.

EXAMPLE 3 Multiply.
$-3(-8x + 3)$

EXAMPLES 4–7 Multiply.
$2(-4y + 7)$

$-9(-4x - 6)$

$-(x - 2)$

$5(3x + 2y - 6z)$

CONCEPT CHECK:

Write a situation that is appropriate to use the Distributive Property for.

GUIDED EXAMPLES:

1. Simplify using the Distributive Property. $5(3 + 8)$

2. Multiply using the Distributive Property. $8 \cdot 17$

3. Multiply. $-2(x - 3)$

4. Simplify using the Distributive Property. $4(x + 3y - 7z)$

Name: _____   Date: _____

Instructor: _____   Section: _____

Notebook 13.6
Multiplication Properties of Real Numbers

What does the Commutative Property of Multiplication say?
_____ when multiplying numbers does not change the result. In symbols, _____.
Give examples:

What does the Associative Property of Multiplication say?
_____ when multiplying numbers does not change the product. In symbols, _____.
Give examples:

What does the Identity Property of Multiplication say?
Multiplying a number by one gives us _____.
In symbols, $1 \cdot a =$ _____ and $a \cdot 1 =$ _____.
Give examples:

EXAMPLES 1–4 Determine which property of multiplication is shown by each equation. The properties are the Commutative Property of Multiplication, the Associative Property of Multiplication, or the Identity Property of Multiplication.

$(-17)(1) = -17$ _____

$(-4 \cdot 2) \cdot 3 = -4 \cdot (2 \cdot 3)$ _____

$(6.5)(-2) = (-2)(6.5)$ _____

$(2)(3 \cdot 6) = (2 \cdot 3)(6)$ _____

Why would you want to change the order of factors or change the grouping of the numbers?

EXAMPLE 5 Multiply.
$\left(\dfrac{1}{4}\right)(-5)(-4)(3)$

203

CONCEPT CHECK:
Which property of multiplication discusses changing the order?

GUIDED EXAMPLES:
Find each product and determine which property of multiplication is used.
1. $12 \cdot 41 = 41 \cdot 12$

2. $(-4 \cdot 5) \cdot 7 = -4 \cdot (5 \cdot 7)$

3. Rewrite using the indicated property, and simplify. $(-3x)(-5)$

4. Multiply. $\left(-\dfrac{1}{2}\right)\left(\dfrac{4}{9}\right)(2)(-9)$

204

Notebook 13.7

Applications of Multiplying and Dividing Real Numbers

Let's review multiplication

Write the rule for multiplying two numbers with the same sign.

Write the rule for multiplying two numbers with different signs.

EXAMPLES 1–4 Multiply the following.

$(-4)(-8)$ $(6.1)(-2)$

$-\dfrac{2}{5} \cdot \dfrac{3}{4}$ $(-4)(5)(-3)$

Write the rule for dividing signed numbers.

EXAMPLES 5–8 Divide the following.

$-21 \div 3$ $-64.8 \div (-8)$

$\dfrac{8}{3} \div 1\dfrac{5}{9}$ $102 \div (-12)$

EXAMPLE 9 An airplane flying to Knoxville was descending at a rate of 2750 feet per minute. If the plane continues at this rate, what is the overall change in altitude after 8 minutes?

Understand the problem.

Create a plan.

Find the answer.

Check the answer.

EXAMPLE 10 During the past year, the financial records for SW Company showed a loss of $6000. Assuming the loss was steady throughout the year, what was the average loss each month?

Understand the problem.

Create a plan.

Find the answer.

Check the answer.

EXAMPLE 11 Cade is snowboarding down a mountain at a rate of 63 meters every 7 seconds. What is the average change of elevation per second? That is, how much does Cade's elevation change every second, on average?
Understand the problem.
Create a plan.
Find the answer.
Check the answer.

EXAMPLE 12 Paul is playing golf with his wife. Paul's golf scores for the first several holes are -2, +2, -3, -1, -2, and 0. What was his average score after these first six holes?
Understand the problem.
Create a plan.
Find the answer.
Check the answer.

CONCEPT CHECK:
A company lost $4 million on a product launch. What expression represents how much the company will lose if it loses the same amount for 5 years?

GUIDED EXAMPLES:

1. Whiteface Mountain has a vertical drop of 3430 feet. Assuming an Olympic skier can ski down the mountain in 28 seconds, what is the average change in vertical drop per second?

2. On January 1, Callie weighed 182 pounds. One month later she weighed 177. Find the overall change in Callie's weight if she continued to lose at this rate for 4 months.

3. At the end of each work day, a cashier must balance his or her register and record any discrepancies. A new cashier has daily discrepancies of $4.14, −$3.27, −$2.99, $0.19 and −$1.12. What is the average daily discrepancy for this cashier for the week?

4. A hot air balloon is descending at a rate of 130 feet per minute. If the balloon continues at this rate, what is the overall change in altitude after 12 minutes?

Notebook 14.1
Introduction to Expressions

How long would three 4-foot-long pieces of rope be if they were laid end-to-end?

How long would three 8-foot-long pieces of rope be, laid end-to-end?

What about three 11-foot-long pieces of rope?

What is a variable?

If the length of each piece of rope is unknown, how would you express the length of three pieces of this rope laid end-to-end?

What if one piece of rope was 2-feet-long and another piece of unknown length?

What are two examples of algebraic expressions used to represent the total length of rope pieces?

An _____ is a combination of _____ and _____, _____ and grouping symbols.

List some examples of algebraic expressions:

A _____ is any number, _____, or _____ of numbers
and/or variables. How are terms separated?

True/False The sign in front of the term is considered part of the term.

EXAMPLES 1–5 Identify the terms in each expression.
$3x$

$2x + 1$

$-7xy^2 + 3z - 2$

60

$-6x - 4z$

207

What are like terms?

Like terms have exactly the same _____ _____.

EXAMPLES 6–10 Identify the like terms in each expression.

$3x + 4y - 7z + 5z$

$2x^2 + 3x$

$-4x + 5 - 2x - 7$

$2x^2 + 3x - 1 + 6x^2 - 9x + 8$

$8a - 7b + 3b + 2b + a$

CONCEPT CHECK:

For the algebraic expression $-5y^2 + 6x - 3xy + 2$, is $6x$ a term in the expression?

GUIDED EXAMPLES:

Identify the terms in each expression.

1. $6y$

2. $-5ab^2 + 2c - 6$

Identify the like terms in each expression.

3. $3a + 5b - 9a + 3c$

4. $7x + 3y - 2x + 9y - 4$

Notebook 14.2
Evaluating Algebraic Expressions

OTHER NOTES

How do you find the perimeter of a square?

What expression is used to represent this?

Find the perimeter of a square with a side of 27 feet.

What does the value of an expression depend on?

Write the steps for evaluating an algebraic expression.
1.

2.

EXAMPLE 1 Evaluate $3x + 6$ for $x = 4$.
Substitute the given value.
Simplify.

EXAMPLE 2 Evaluate $5x^2$ for $x = 6$.
Substitute the given value.
Simplify.

EXAMPLE 3 Evaluate $3y^2 - y$ for $y = -7$.
Substitute the given value.
Simplify.

EXAMPLE 4 Evaluate $\dfrac{a + 2b}{4c}$ for $a = 0$, $b = -6$, and $c = 2$.

Substitute the given values.
Simplify.

Note: Use the Distributive Property to rewrite $2(l + w)$.

EXAMPLE 5 The perimeter of a rectangle can be found by the expression $2(\underline{l} + w)$ or $2l + 2w$, where l is the length and w is the width. Find the perimeter of this rectangle if $l = 5.4$ and $w = 8$. You will need to label the rectangle.)

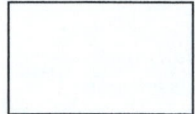

CONCEPT CHECK:
When evaluating an algebraic expression, the value of the expression depends on the value of the _____.

GUIDED EXAMPLES:
Evaluate for the given values.

1. $7m + 2$ for $m = 1$

2. $\dfrac{3a - 4b}{2c}$ for $a = 9$, $b = 7$, and $c = -3$

3. $9d^2$ for $d = 4$

4. $6w^2 + 3w$ for $w = -5$

Name: _____   Date: _____

Instructor: _____   Section: _____

Notebook 14.3
Simplifying Expressions

An _____ is a combination of _____ and
_____, operation symbols and _____.

What would you get if you combined 5 inches and 6 inches?

Could you combine 4 inches and 3 square feet?   Why?

How do you simplify an expression?

EXAMPLES 1–4 Combine like terms.
$13x - 2x$

$-4a + 3b + 9a$

$2x^3 + 9x^2 - x + 7$

$\dfrac{4}{3}m + \dfrac{2}{3}m - m$

Why might it be helpful to rearrange the terms when simplifying expressions?

Caution! The sign _____ of the term stays with the term.

EXAMPLE 5 Combine like terms. $2a + 32b - 25c - 12b + 15a + 13c$
Rearrange the terms.

Combine the like terms.

EXAMPLE 6 Combine like terms. $15x^3 + 2 - 8x^2 + x^3 - 9x + 13x^2 - 3$
Rearrange the terms.

Combine the like terms.

EXAMPLE 7 Combine like terms. $16x^2y - 3xy^2 + 5xy - 2x^2y - 4xy^2$
Rearrange the terms.

Combine the like terms.

CONCEPT CHECK:
What does it mean to combine like terms?

GUIDED EXAMPLES:
Combine like terms.
1. $23x + 6x$

2. $6x^2 + 18y - 4x^2 + 12y$

3. $6a^2 + 14a - 19a^2 - 34b - 8b + 4a + 15b + a$

4. $12ab^2 + 6.5ab - 5a^2b - 4ab^2 + ab$

Notebook 14.4
Simplifying Expressions with Parentheses

What are some types of grouping symbols used in expressions?

What does it mean if an algebraic expression is simplified?

If an algebraic expression contains parentheses, what can be done to remove parentheses?

EXAMPLE 1 Simplify.
$5 + 4(a + b)$

What does a negative sign in front of the parentheses mean?

How do you simplify an expression with a negative sign in front of the parentheses?

EXAMPLES 2 & 3 Simplify.
$-(4x - 3y)$

$-(5 + 6a)$

Write the steps for simplifying algebraic expressions.
1.

2.

3.

EXAMPLES 4 & 5 Simplify.
$8x - 3(4x - 5)$

$2[3(9x - 8)]$

EXAMPLE 6 Simplify.

$14x - 2[3x + 3(5)]$

_____ are also considered grouping symbols.

EXAMPLE 7 Simplify. $\dfrac{7-(4-x)}{3+2(x-5)}$

Simplify the numerator.

Simplify the denominator.

CONCEPT CHECK:

How are parentheses removed?

GUIDED EXAMPLES:

Simplify.

1. $7 + 2(m + n)$

2. $-(7a + 6b)$

3. $7x + 3[2x + 2(6x - 1)]$

4. $\dfrac{4(3x - 14)}{9(5y + 2)}$

214

Name: _____   Date: _____

Instructor: _____   Section: _____

Notebook 14.5
Translating Words into Symbols

What does the Commutative Property of Addition state?

What does the Commutative Property of Multiplication state?

True/False The order in which you add or multiply two numbers does not matter.

List some key words (phrases) for addition representing "$x + 4$"

The _____ of a number and four   A number _____ by four
Four _____ a number   Four is _____ a number
Four _____ a number   A number _____ four.

List some key words (phrases) for multiplication representing "$2x$":
_____ a number   _____ a number
The _____ of two and a number   Two _____ a number
Two _____ a number

EXAMPLES 1–3 Translate into an algebraic expression.
The sum of 8 and a number

Triple a number

75% of a number

List some key words (phrases) for subtraction representing "$a - 6$"
The _____ a number and six   Six is _____ a number
A number _____ by six   Six _____ a number
Six _____ a number   A number _____ six

List some key words (phrases) for division representing "$\dfrac{a}{7}$":

A number _____ seven
The _____ of a number and seven
A number _____ seven

Caution! Which operations do not work with the Commutative Property? What does this mean?

EXAMPLES 4–7 Translate into an algebraic expression.
Five less than twelve

Twelve less than five

Fifty divided by one

One divided by fifty

EXAMPLES 8–10 Translate into an algebraic expression. Use n for the variable.
Twelve less than a number n

The difference between –9 and a number n

The quotient of a number n and 2.1

EXAMPLES 11–13 Translate into an algebraic expression. Use parentheses if necessary.
Seven more than double a number x

A number x is tripled and then increased by 8

One-half of the sum of a number and 3

EXAMPLE 14 Use an expression to describe the measure of each angle. The figure is not drawn to scale. The measure of the second angle of a triangle is double the measure of the first angle, and the third angle is 15° more than the measure of the second angle. You will need to draw the triangle.

CONCEPT CHECK:
What phrase best represents $2(n - 3)$?

GUIDED EXAMPLES:
Translate into an algebraic expression.
1. A number increased by thirty-eight

2. Fifteen less than a number

3. The quotient of a number and 5.5

4. A number x out of sixty-two

Name: _____   Date: _____

Instructor: _____   Section: _____

Notebook 15.1
Translating Words into Equations

A _____ is a letter or symbol that is used to represent an unknown quantity.

What are the basic steps for understanding word problems.
a.

b.

c.

What is an equation?
What is it that all equations contain?

What are the steps for writing an equation from a word problem?
a.

b.

c.

d.

e.

EXAMPLE 1 Translate the following into an equation. Let n represent the number.
One-third of a number is fourteen.

EXAMPLE 2 Translate the following into an equation. Do not solve.
Five more than six times a number is three hundred five.

EXAMPLE 3 Translate the following into an equation. Do not solve.
The larger of two numbers is three more than twice the smaller number.
The sum of the numbers is thirty-nine.

Understand the problem.

Write the equation.

EXAMPLE 4 Translate the following into an equation. Do not solve.
The annual snowfall in Juneau, Alaska, is 105.8 inches. This is 20.2 inches less than three times the annual snowfall in Boston, Massachusetts.

Understand the problem.

Write the equation.

CONCEPT CHECK:

Is $\dfrac{28x}{9}$ an equation?

GUIDED EXAMPLES:
Translate into an equation. Do not solve.

1. One-fourth of a number is eleven.

2. Six more than three times a number is fifty-seven.

3. Four times a number decreased by thirteen is thirty-five.

4. The larger of two numbers is eleven more than three times the smaller number. The sum of the numbers is forty-two.

218

Notebook 15.2
Linear Equations and Solutions

An _____ is a mathematical statement that two _____ are
_____.

What are some examples of equations?

What is a letter or symbol that represents an unknown quantity?

Example "Five more than a number is equal to eight."

What is a solution of an equation?

What is a solution of $x + 5 = 8$?   How do you know?

How many solutions does an equation have?

Write the steps for determining if a given value is a solution.
1.

2.

3.

EXAMPLE 1 Is 2 a solution of the equation $3x - 1 = 5$?
Substitute.

Simplify.

True?

EXAMPLES 2 & 3
Is -1 a solution of the equation $2x + 6 = -1$?

Is -3 a solution to $7x - 2 = 5$?

Define a linear equation in one variable.

EXAMPLES 4–6 Determine if each equation is a linear equation.

$2x + 3 = 1$

$2x = 5$

$6x^2 - 3 = 4$

CONCEPT CHECK:
Why is 0 not a solution to the equation $2x + 1 = 5$?

GUIDED EXAMPLES:

1. Is 9 a solution of the equation $4x - 11 = 25$?

2. Is -5 a solution of the equation $4 = 7x - 2$?

3. Is -4 a solution of the equation $5x + 3 = -17$?

4. Determine if $2x - 9 = -15$ is a linear equation.

220

Notebook 15.3
Using the Addition Property of Equality

OTHER NOTES

A _____ of an equation is the number(s) that, when substituted for the
_____, makes the equation _____.

To determine if a given value is a _____ of an equation,
_____ this value into the equation and _____ each side
according to the order of operations.

What are equivalent equations?

When an equation is in the form $x =$ some number, that number is the
_____ of the equation.

What is the process of finding the solution(s) of an equation called?

What is the goal of solving an equation?

The Addition Property of Equality states that if the _____
number is added to _____ of an equation, the results
on both sides are _____ in value. That is, adding the _____
_____ to both sides of an equation does not change the _____.

How is this represented in symbols?

Does this property work for subtraction as well?   Why or why not?

Which property "undoes" addition?
Which property "undoes" subtraction?

EXAMPLE 1 Solve for x. Check your solution.
$x + 16 = 20$

Write the steps for solving an equation using the addition property.

1.

2.

3.

EXAMPLE 2 Solve for x. Check your solution.

$-14 = x - 3$

EXAMPLE 3 Solve for x. Check your solution.

$x - \dfrac{1}{2} = -\dfrac{5}{2}$

When solving an equation, _____ both sides of the equation whenever possible. _____ will make it easier to work with.

EXAMPLE 4 Solve for x. Check your solution.

$15 + 2 = 3 + x + 6$

CONCEPT CHECK:
What would the first calculation be in order to solve the equation $2 - 8 = 2x + 9$?

GUIDED EXAMPLES:
Solve for x. Check your solution.

1. $\dfrac{8}{5} = x - \dfrac{2}{5}$

2. $x + 0.6 = -8.1$

3. $5 + x - 3 = -2 + 7$

4. $27 + x - 14 = -5 + 39$

Notebook 15.4
Using the Multiplication Property of Equality

How does the Addition Property of Equality help to solve an equation?

The Multiplication Property of Equality states that if both sides of an
_____ are _____ by the _____ number, the
_____ does not change.

How is this represented in symbols?

Does this property work for division as well?   Why or why not?

Caution! We cannot divide by _____. So we must restrict our divisors to
_____ numbers.

_____ and _____ are reverse operations.

EXAMPLE 1 Solve for x. Check your solution.
$$\frac{x}{3} = -15$$

Write the steps for solving an equation using the multiplication property.
1.

2.

3.

EXAMPLE 2 Solve for x. Check your solution.
$3x = 21$

223

EXAMPLE 3 Solve for *x*. Check your solution.

$20 = -4x$

When solving an equation, _____ both sides of the equation whenever possible. _____ will make it easier to work with.

EXAMPLE 4 Solve for *x*. Check your solution.

$2x - 5x = -12$

CONCEPT CHECK:

What is the reverse operation of multiplication?

GUIDED EXAMPLES:

Solve for *x*. Check your solution.

1. $-18 = 6x$

2. $-5x = 35$

3. $\dfrac{x}{10} = -6$

4. $8x - 9x = 10$

Notebook 15.5
Using the Addition and Multiplication Properties Together

Jenny scored several goals in field hockey during April. Her teammates scored three more than five times the number of goals she scored. Her teammates scored 18 goals. How many goals did Jenny score?

What equation do we need to solve?

The properties of equality state that you can _____ (or _____)
and _____ (or _____) both sides of an equation
by the_____ without changing the _____.

EXAMPLE 1 Solve for x to determine how many goals Jenny scored, and then check your solution. Be sure to show all steps.
$5x + 3 = 18$

Write the steps for solving an equation of the form $ax + b = c$.
1. Get the _____ alone on one side of the equation.

2. Get the _____ alone on one side of the equation.

3. Simplify any _____, if needed.

4. _____ your solution.

EXAMPLE 2 Solve for x. Show all steps.
$$-\frac{1}{2}x + 10 = 16$$

EXAMPLE 3 Solve for x. Then check. Show all steps.

$6x - 8 = -2$

EXAMPLE 4 Solve for x. Then check. Show all steps.

$4 = -7 + 8x$

CONCEPT CHECK:

Erica and Stefan played a video game. Erica scored 8 less than 4 times Stefan's score. Erica's score was 1000 points. Let x = the number of points Stefan scored. Write the equation that finds the number of points Stefan scored.

GUIDED EXAMPLES:

Solve for x. Check your solution.

1. $9x + 1 = 28$

2. $\dfrac{1}{8}x - 6 = -10$

3. $-15 = 2x - 7$

4. $3x - 5.6 = 12.4$

Name: _____   Date: _____

Instructor: _____   Section: _____

Notebook 16.1
Solving Equations with Variables on Both Sides

A _____ of an equation is the number(s) that, when substituted for the _____, makes the equation true.

The process of finding the solution(s) to an equation is called
_____.

What is the goal of solving an equation?

The properties of equality state that you can _____ (or _____) and _____ (or _____) both sides of an equation by the_____ without changing the _____.

How do you know whether to add or subtract to get the variable terms on one side of the equation?

EXAMPLE 1 Solve for x. Check your answer.
$9x = 6x + 15$
Get the variable terms on one side.

Get the number terms on the other side.

Get the variable alone on one side.

Simplify.

EXAMPLE 2 Solve for x. Check your answer. Show all steps.
$9x + 4 = 7x - 2$
Get the variable terms on one side.

Get the number terms on the other side.

Get the variable alone on one side.

Simplify.

You can simplify one or both sides of the equation by _____ like terms that are on the _____ of the equation.

227

EXAMPLE 3 Solve for x. Check your answer. Show all steps.

$5x + 26 - 6 = 9x + 12x$

Simplify each side.

Get the variable terms on one side.

Get the number terms on the other side.

Get the variable alone on one side.

Simplify.

EXAMPLE 4 Solve for x. Check your answer. Show all steps.

$-x + 8 - x = 3x + 10 - 3$

Simplify each side.

Get the variable terms on one side.

Get the number terms on the other side.

Get the variable alone on one side.

Simplify.

CONCEPT CHECK:

What is the first goal when solving an equation with a variable m on both sides?

GUIDED EXAMPLES:

Solve for x.

1. $5x + 8 = 10x$

2. $7x - 7 = 12x + 8$

3. $11x + 6x = -20 + 13x - 12$

4. $-13x + 3 + 5x = 1 - 6x + 16$

Name: _____   Date: _____

Instructor: _____   Section: _____

Notebook 16.2
Solving Equations with Parentheses

Recall: For all real numbers a, b, and c,
$a(b + c) = $ _____.(Distributive Property)

In order to solve an equation with parentheses, _____
first using the _____to remove parentheses.

EXAMPLE 1 Solve for x. Check your answer. Show all steps.
$2(x + 5) = -12$
Simplify each side.
Get the variable terms on one side.
Get the number terms on the other side
Get the variable alone on one side.
Simplify.

EXAMPLE 2 Solve for x. Check your answer. Show all steps.
$-5(x - 3) + 7 = x - 8$
Simplify each side.
Get the variable terms on one side.
Get the number terms on the other side
Get the variable alone on one side.
Simplify.

EXAMPLE 3 Solve for x. Check your answer. Show all steps.
$5(x + 1) - 3(x - 3) = 17$
Simplify each side.
Get the variable terms on one side.
Get the number terms on the other side.
Get the variable alone on one side.
Simplify.

EXAMPLE 4 Solve for x. Check your answer. Show all steps.
$3(-x - 7) = -2(2x + 5)$
Simplify each side.
Get the variable terms on one side.
Get the number terms on the other side.
Get the variable alone on one side.
Simplify.

229

EXAMPLE 5 Solve for x. Check your answer. Show all steps.

$3(0.5x - 4.2) = 0.6(x - 12)$

Simplify each side.

Get the variable terms on one side

Get the number terms on the other side.

Get the variable alone on one side.

Simplify.

EXAMPLE 6 Solve for x. Check your answer. Show all steps.

$2(18x - 5) + 2 = 24x - 3(12x + 8)$

Simplify each side.

Get the variable terms on one side.

Get the number terms on the other side.

Get the variable alone on one side.

Simplify.

CONCEPT CHECK:

What is the correct application of the Distributive Property for

$5(x - 3) - 2(x + 1) = 15$?

GUIDED EXAMPLES:

Solve for x.

1. $3(x - 7) = 6$

2. $5(-2x + 1) = -7(x - 2)$

3. $4(-2x - 3) = -5(x - 2) + 2$

4. $5(10x - 1) + 7 = 35x - 4(5x + 3)$

Name: _____   Date: _____

Instructor: _____   Section: _____

Notebook 16.3
Solving Equations with Fractions

The _____ (LCD) of two or more fractions is
the _____ (LCM) of the _____ of the
fractions.

True/False The equation-solving procedures are the same for equations with or
without fractions.

When solving equations with fractions, we can perform an extra step that will
make the calculations a little "nicer". What is this extra step?

EXAMPLE 1 Solve for x. Check your answer. Show all steps.
$$\frac{1}{4}x - \frac{2}{3} = \frac{5}{12}x$$

Find the LCD of the fractions.
Multiply both sides by the LCD.
Use the Distributive Property.
Solve.

Note: Multiplying both sides of the equation by the LCD and using
_____ is the same as _____ each term
in the equation by the _____.

EXAMPLE 2 Solve for x. Check your answer. Show all steps.
$$\frac{x}{3} + 3 = \frac{x}{5} - \frac{1}{3}$$

Find the LCD of the fractions.
Multiply each term by the LCD.
Solve.

EXAMPLE 3 Solve for x. Check your answer. Show all steps.
$$\frac{x+5}{7} = \frac{x}{4} + \frac{1}{2}$$

Find the LCD of the fractions.

Multiply each term by the LCD.

Solve.

How can you use this to solve equations containing decimals?

EXAMPLE 4 Solve for x. Check your answer. Show all steps.

$0.6x - 1.3 = 4.1$

Multiply each term by 10.

Solve.

Why did we choose to multiply by 10?
How do you know what to multiply by?

CONCEPT CHECK: To make the process of solving $\dfrac{1}{2}x - 4x = \dfrac{1}{3}x$ easier,

by what number can both sides of equation be multiplied?

GUIDED EXAMPLES:
Solve for x.

1. $\dfrac{2}{3}x + \dfrac{1}{2} = -\dfrac{5}{6}$

2. $2 - \dfrac{x}{10} = -\dfrac{x}{4} + \dfrac{1}{2}$

3. $\dfrac{x+6}{9} = \dfrac{x}{6} + \dfrac{1}{2}$

4. $-2.7x + 6.4 = -4(3x - 1.6)$

Name: _____   Date: _____

Instructor: _____   Section: _____

Notebook 16.4
Solving a Variety of Equations

Write the steps for solving an equation:

1. Remove _____ by using _____.

2. If _____ or _____ remain, multiply each
   term by the _____ of all fractions.

3. _____ each side if possible.

4. _____ or _____ terms on both sides to get all
   _____ on one side of the equation.

5. _____ or _____ number terms on both sides to get all
   _____ on the OTHER side of the equation.

6. _____ or _____ both sides of the equation to get the
   _____ alone on one side of the equation.

7. _____ the solution.

8. _____ your solution.

True/False Every step of the procedure is needed in each problem.

EXAMPLE 1 Solve for x. Check your answer. Show all steps.
$3(6x - 4) = 4(3x + 9)$

EXAMPLE 2 Solve for x. Check your answer. Show all steps.
$2(x + 1) = 5(x - 2) + 3$

EXAMPLE 3 Solve for x. Check your answer. Show all steps.

$\frac{1}{3}(x - 2) = \frac{1}{5}(x + 4) + 2$

What does it mean if an equation has no solution?

What symbol is used for no solution?

What is a contradiction?

EXAMPLE 4 Solve for x. Check your answer. Show all steps.

$5(x + 3) = 2x - 8 + 3x$

What is an identity?

What does it mean if an equation has an infinite number of solutions?

EXAMPLE 5 Solve for x. Check your answer. Show all steps.

$9x - 8x - 7 = 3 + x - 10$

CONCEPT CHECK:

When the equation $2x - x + 1 = x + 3 - 2$ is solved, the result is $1 = 1$.
True/False The equation is an identity.

GUIDED EXAMPLES:

Solve for x.

1. $\dfrac{1}{5}(5x - 10) = \dfrac{1}{3(9x + 6)}$

2. $\dfrac{1}{2}(x + 5) = \dfrac{1}{5}(x - 2) + \dfrac{1}{2}$

3. $-1 + 5(x - 2) = 12x + 3 - 7x$

4. $9(x + 3) - 6 = 24 - 2x - 3 + 11x$

234

Name: _____   Date: _____

Instructor: _____   Section: _____

Notebook 16.5
Solving Equations and Formulas for a Variable

OTHER NOTES

What is a formula?

Write some examples for formulas.

What is the procedure for solving a formula for a specified formula?

1. Identify the _____.

2. Remove any _____.

3. If fractions or decimals remain, _____ each term by the _____ of all fractions.

4. _____ each side if possible.

5. _____ terms on both sides to get all terms containing the specified variable on _____ of the equation.

6. _____ terms on both sides of the equation to get all terms NOT containing the _____ on the other side.

7. _____ both sides of the equation to get the specified _____ alone on one side of the equation.

What is the hint for solving a formula for a specified variable, treat…

EXAMPLE 1 Solve $5x + 2 = 17$ and $ax + b = c$ for x.

EXAMPLE 2 Solve for t. $d = rt$

235

EXAMPLES 3–6 Solve for the specified variable.

$C = 2\pi r$, solve for r

$a = \dfrac{v}{t}$, solve for v

$x + y = 8$, solve for x

$a - b = -3$, solve for a

EXAMPLES 7 & 8 Solve for the specified variable.

$5x + 3y = 6$, solve for y

$y = mx + b$, solve for x

CONCEPT CHECK:

True/False $P = 2L + 2W$ can be solved for W using only the Multiplication Property of Equality.

GUIDED EXAMPLES:

1. Solve for r. $I = prt$

2. Solve for l. $P = 2l + 2w$

3. Solve for x. $x + 3y = 7$

4. Solve for y. $5x - 2y = -8$

236

Name: _____   Date: _____

Instructor: _____   Section: _____

Notebook 16.6
Solving and Graphing Linear Inequalities

Review of Inequalities

A _____ is a statement that shows the relationship between any two real numbers that are _____ _____.

Fill in the table:

Symbol	Phrase
$<$	
$>$	
$\leq$	
$\geq$	

What is a linear inequality?
Examples of linear inequalities:

What is a solution of an inequality?

What does the inequality $x > 3$ mean?

What is the graph of an inequality?

Write the steps for graphing a linear inequality.
1. Plot the _____.
      a.
      b.
2. Shade _____ …

EXAMPLES 1 & 2 Graph each inequality on a number line.

$x > 3$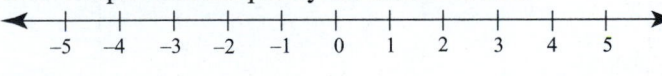

$x \leq -1$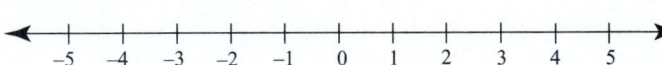

When do you use an open circle on the graph and when do you use a closed circle on a graph?

What does it mean to solve an inequality?

What happens to an inequality when you add 2, subtract 2, multiply by 2 and divide by 2 on both sides of an inequality?

What happens when you add -2 or subtract -2 to both sides?

What happens when you multiply or divide both sides of an inequality by a negative number?

To solve a linear inequality, use the same procedure used to solve an equation, except, the _____ of an inequality must be _____ if you _____ or _____ both sides by a _____.

EXAMPLE 3 Solve and graph the inequality. $5x + 2 < 12$

EXAMPLE 4 Solve and graph the inequality. $5 - 4x \geq -7$

CONCEPT CHECK:
What is the word phrase that translates to $x \geq 3$?

GUIDED EXAMPLES:
Graph the inequality on a number line.

1. $x > 5$
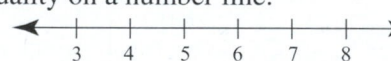

2. $-2x - 6 < 0$
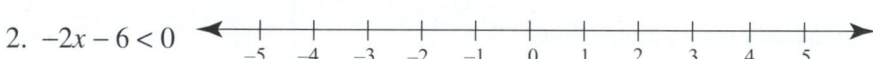

Solve and graph the inequality.

3. $7 + 4x \leq 5$

4. $2(x - 9) \geq -4(x - 6)$
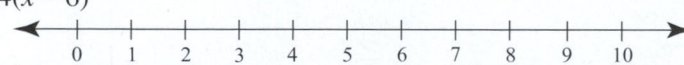

238

Name: ______________________________ Date: ________________

Instructor: ____________________________ Section: ______________

Notebook 16.7
Applications of Linear Equations and Inequalities

OTHER NOTES

What are the steps for writing an equation from a word problem?

a.

b.

c.

d.

e.

EXAMPLE 1 Translate the following into an equation and solve.
One-third of the length of a race is nineteen miles. Find the length of the race.

EXAMPLE 2 Translate the following into an equation and solve.
Five more than six times a number is one hundred eighty-five. Find the number.

EXAMPLE 3 Translate the following into an equation and solve.
A mother is seven years older than twice the age of her daughter. The sum of their ages is forty-three. Find their ages.

EXAMPLE 4 Translate the following into an equation and solve.
The largest Bluefin tuna caught in the Atlantic Ocean weighed 1496 pounds. This is 246.5 pounds less than five times the weight of the largest Bluefin tuna caught in the Pacific Ocean. Find the weight of the largest tuna caught in the Pacific.

EXAMPLE 5 Translate the following into an equation and solve.
Heather makes $14.50 per hour as a lifeguard. Her paycheck is calculated by multiplying $14.50 times the number of hours she worked. How many hours will she have to work to earn more than $522?

CONCEPT CHECK:
Would this problem be solved with an equation or an inequality sign? Two minor league baseball players had a total of 260 hits. Domingo had 16 more hits than Brown. Find the number of hits for each player.

GUIDED EXAMPLES:
Translate the following into an equation and solve.
1. One-sixth of a number is forty-seven.

2. Eight more than three times a number is seventy-seven.

3. The larger of two numbers is sixteen more than three times the smaller number. The sum of the numbers is three hundred eighty-eight. Find both numbers.

4. Jeff needs 13 pounds of meat for a cookout. What is the maximum cost per pound at which Jeff can buy the meat if he wants to spend at most $60? Round your answer to the nearest cent.

Name: _____   Date: _____

Instructor: _____   Section: _____

Notebook 17.1
The Rectangular Coordinate System

What makes up a rectangular coordinate system?

What is the point at which these number lines meet?

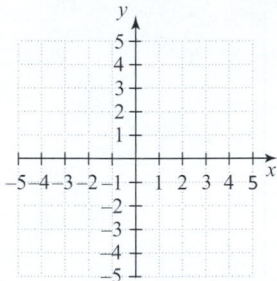

On a graph like this, in which direction is the *x*-axis positive?
Which direction is positive on the *y*-axis?
What are the four sections of the graph called?
Label them on the graph above.

True/False The order is very important when writing an ordered pair.
Write an ordered pair using an *x*-value of *x*, and a *y*-value of *y*.

What is the *x*-coordinate of an ordered pair?
What is the *y*-coordinate of an ordered pair?

What is the ordered pair for the origin?

EXAMPLE 1 Plot the points (4, 2), (3, −2), (−3, 3), and (−1, −4). Label them *A*, *B*, *C* and *D*, respectively.

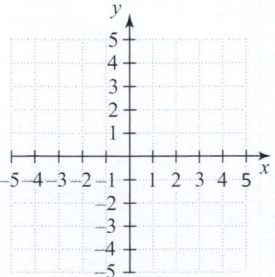

Note: If a point lies on an axis, is it in a quadrant?

241

EXAMPLE 2 Plot the points on the graph. Identify which quadrant each point lies in. Label the points *A*, *B*, *C* and *D*, respectively.

A(4, −5)

B(−5, 4)

C(3, 0)

D(2, 2)

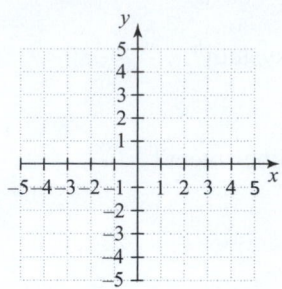

EXAMPLE 3 Find the coordinates of the indicated points. Write each point as an ordered pair. Identify which quadrant each point lies in.

A

B

C

D

E

F

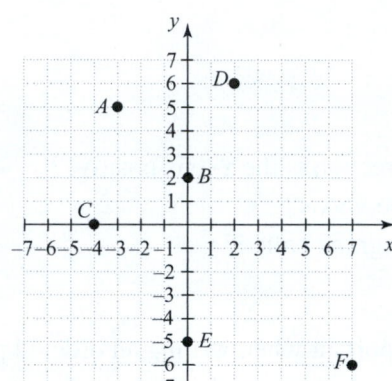

CONCEPT CHECK:

For the ordered pair (3, −4), in which direction would the coordinate −4 be plotted, and by how many units?

GUIDED EXAMPLES:

Plot the given points on the graph. Identify which quadrant each point lies in.

1. (3, 1)   2. (−2, −5)   3. (−4, 4)   4. (0, −3)

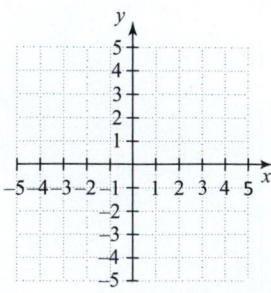

242

Name: _____ Date: _____

Instructor: _____ Section: _____

Notebook 17.2
Graphing Linear Equations by Plotting Points

What is the standard form of a linear equation?

What do A, B and C represent?

How can you tell if an ordered pair is a solution to an equation?

EXAMPLE 1 Determine whether $(3, 5)$ is a solution to the equation $3x + 2y = 19$.

EXAMPLE 2 Determine whether $(6, -4)$ is a solution to the equation $3x = 2y + 10$.

How many solutions does a linear equation in two variables have?

EXAMPLE 3 Find five solutions to $x + y = 10$.

$10 + 0 = 10$	$x = 10$	$y = 0$	is a solution	The ordered pair is $(10,0)$

Write the steps for finding a solution to a linear equation.
1.

2.

EXAMPLE 4 Find three solutions to $2x + y = 13$.
Substitute a value.
Solve for the other variable.
Repeat.

Where do we get the values for x?

OTHER NOTES

EXAMPLE 5 Find three solutions to $y = \dfrac{1}{3}x - 2$.

Substitute a value.
Solve for the other variable.
Repeat.

Why did we choose these particular values (0, 3 and 6) of x?

Why use a table of values?

EXAMPLE 6 Find three solutions to $x + y = -4$.

x	y

Write the steps for graphing a linear equation by plotting points.
1.

2.

3.

EXAMPLE 7 Graph the equation $y = 2x + 1$.

Find three ordered pairs.
Plot the ordered pairs on a graph.
Draw a line.

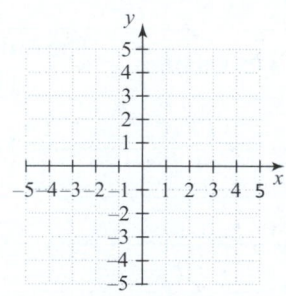

CONCEPT CHECK:
True/ False   A solution to the equation $x - y = 5$ is (9, 4).

GUIDED EXAMPLES:
Determine whether the given point is a solution to the equation.
1. $2x + 6y = 50$; (4, 7)

2. $4x - 5y = 18$; (3, – 1)

Find two solutions to the given equation.
3. $3x + y = 8$

4. $x + y = -9$

244

Name: _____   Date: _____

Instructor: _____   Section: _____

Notebook 17.3
Graphing Linear Equations Using Intercepts

What is an intercept?

The _____ of a line is the point where the line crosses the _____. The ordered pair of an x-intercept is _____.

How do you find the x-intercept of a linear equation?

The _____ of a line is the point where the line crosses the _____. The ordered pair of a y-intercept is _____.

How do you find the y-intercept?

Note: The graph of a linear equation is a _____.

EXAMPLE 1 Find the x-intercept and y-intercept of $3x - 6y = 12$.
x-intercept: y-intercept:

EXAMPLE 2 Find the x-intercept and y-intercept of $y = \dfrac{4}{5}x$.

x-intercept: y-intercept:

Note: What does it mean if the line passes through the origin?

Write the steps for graphing a linear equation using the intercepts.
1.

2.

3.

245

EXAMPLE 3 Graph $2x - y = 4$ using the intercepts.

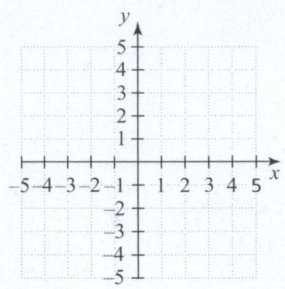

EXAMPLE 4 Graph $y = \dfrac{2}{3}x - 2$ using the intercepts.

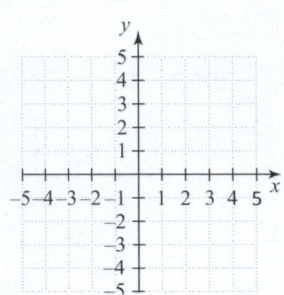

CONCEPT CHECK:
True/False $(2, 0)$ is a y-intercept.

GUIDED EXAMPLES:
Find the x- and y-intercepts of each equation.
1. $5x + 3y = 15$

2. $y = 3x - 7$

3. $4x + y = 8$

4. $y = \dfrac{1}{2}x + 4$

246

Notebook 17.4
Graphing Linear Equations of the form $x = a$, $y = b$, and $y = mx$

Draw a picture that shows one x-intercept and one y-intercept.

What would a line look like that has only one intercept?
Show the three cases.

Lines that pass through the _____ have only one _____.
What will the intercept will be?
Write some examples.

EXAMPLE 1 Graph $6x - 2y = 0$ using three points.

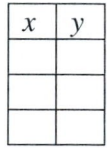

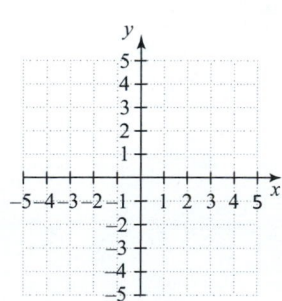

x	y

Note: How can you identify an equation of this type by looking at it?

What type of equation has a graph that is a horizontal line?
Write some examples of these equations.

What is missing from these equations?

EXAMPLES 2 & 3 Graph each equation.

$y = 4$ $3y + 14 = 5$

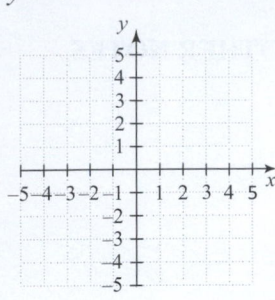   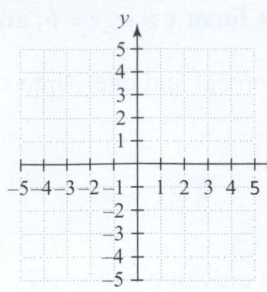

What type of equation has a graph that is a vertical line?
Write some examples of these equations.

What is missing from these equations?

EXAMPLES 4 & 5 Graph each equation.

$x = 3$ $3x + 7 = -5$

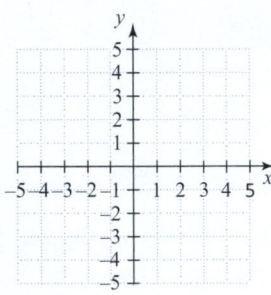   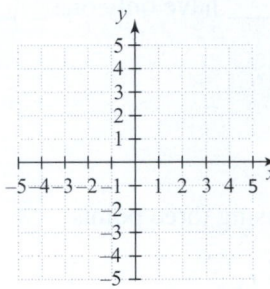

CONCEPT CHECK:
True/False $x + 5 = 0$ has only one y-intercept.

GUIDED EXAMPLES:
Graph each equation.

1. $15x - 5y = 0$ 2. $y - 5 = 0$ 3. $x + 1 = 0$

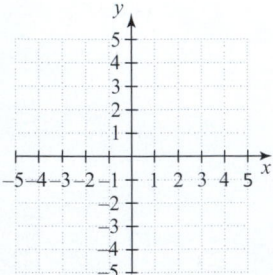   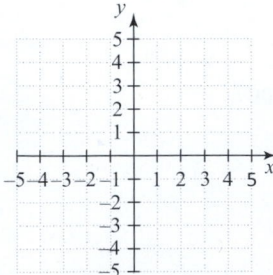   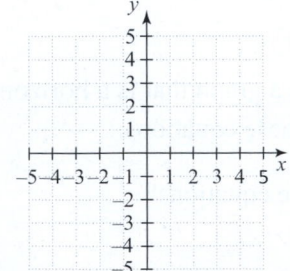

248

Notebook 17.5
Applications of Graphing Linear Equations

An ordered pair is a _____ to an equation if it results in a true statement when the values for the x-coordinate and the y-coordinate are substituted into the _____.

Write the steps for finding a solution to a linear equation in two variables.
1. _____ a value for one of the _____.

2. _____ the equation to find the other variable.

The _____ is an ordered pair with the coordinates _____.
The _____ is an ordered pair with the coordinates _____.

EXAMPLE 1 Paul received a $75 gift card to an electronics store. Used DVDs cost $3 each, and used video games cost $17 each.. The equation that represents buying x DVDs and y games is $3x + 17y = 75$.
What does the x-intercept represent in this equation?

Find the x-intercept.

If Paul buys three games, how many DVDs can he buy?

EXAMPLE 2 The amount of money spent annually on food and beverages in restaurants since 2000 can be approximated by the equation $y = 16.8x + 223$, where y is the sales in billions of dollars and x is the number of years since 2000. What does the y-intercept represent in this equation? $(0, y)$

Find the y-intercept.

What does the point $(3, 273.4)$ represent in this equation?

Find the amount of money spent in restaurants in 2006.

EXAMPLE 3 The amount of money spent annually on food and beverages in restaurants since 2000 can be approximated by the equation $y = 16.8x + 223$, where y is the sales in billions of dollars and x is the number of years since 2000. Graph the equation using the ordered pairs that we found in the previous example:

(0, 223)

(6, 323.8)

(10, 391)

(14, 458.2)

Approximate the amount of money spent in restaurants in 2008.

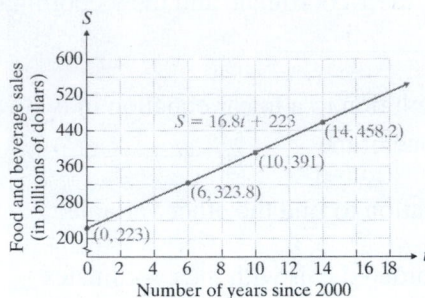

EXAMPLE 4 The number of calories burned by an average person while jogging can be given by the equation $C = \dfrac{28}{3}m$, where m is the number of minutes. Graph the equation using the ordered pairs associated with $m = 0, 15, 30, 45$ and 60.

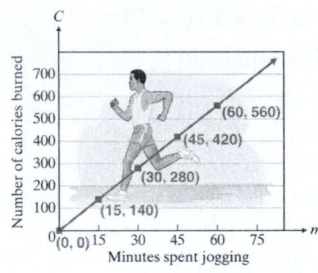

CONCEPT CHECK: The yearly profits of a city's restaurant since 1990 can be approximated by the equation $y = 15.4X + 200$, where y is the sales in thousands of dollars and x is the number of years since 1990. That is the ordered pair solution for the year 2000?

GUIDED EXAMPLES:

Lee wants to buy hats and scarves for a local homeless shelter. The hats cost $8 each, and scarves cost $14 each. Lee has $280 to spend. The equation that represents buying x hats and y scarves is $8x + 14y = 280$.

1. Find the x-intercept, and state what the x-intercept represents in this problem. Then find the number of hats Lee can buy if he buys 16 scarves.

2. Find the y-intercept, and state what the y-intercept represents in this problem. Then find the number of scarves Lee can buy if he buys 21 hats.

3. The number of foreign students enrolled in American colleges in a given year can be approximated by the equation $S = 11t + 395$, where S is the number of students in thousands, and t is the number of years since 2000. Graph the equation using the ordered pairs for $t = 0, t = 4, t = 8,$ and $t = 16$.

250

Name: _____   Date: _____

Instructor: _____   Section: _____

Notebook 18.1
The Slope of a Line

How can we describe the slope of a line?

What letter is used to represent slope?

How do you find the rise of a line?

How do you find the run of a line?

EXAMPLES 1 & 2 Find the slope.

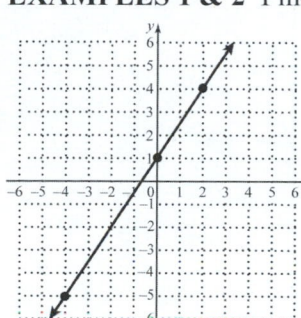   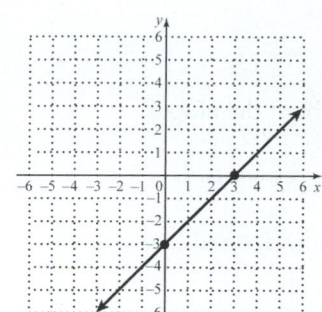

Lines with a positive slope will _____ from left to right.

EXAMPLES 3 & 4 Find the slope.

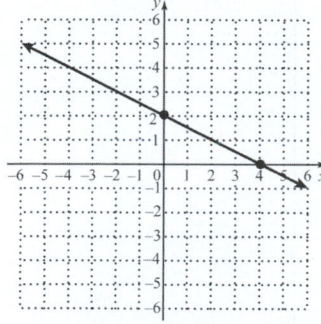   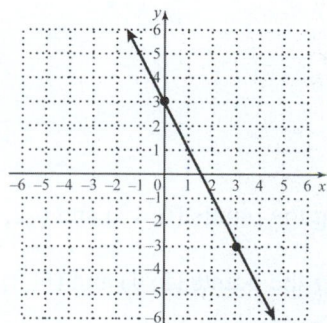

Lines with a negative slope will _____ from left to right.

Define slope.

Write the slope formula for when you know two points (x_1, y_1) and (x_2, y_2).

251

EXAMPLES 5 & 6 Find the slope of the line which contains the given points.

Use the slope formula: $m = \dfrac{y_2 - y_1}{x_2 - x_1}$.

(1, 3) and (5, 11)

(−3, −4) and (7, −9)

On a graph, what does slope measure?

If a line is steeper, what does that say about the slope?

What do you know about the slope of a horizontal line?
How do you know?

What do you know about the slope of a vertical line?
How do you know?

True/False No slope is the same as zero slope.

CONCEPT CHECK:
What is the slope of a horizontal line?

GUIDED EXAMPLES:
1. Find the slope of the line containing the points (6, 13) and (−2, 4).

2. Find the slope of the line containing the points (3, 4) and (6, 8).

3. Find the slope of the line $y = -6$.

4. Find the slope of the line $x = 5.4$.

252

Notebook 18.2
Slope-Intercept Form

OTHER NOTES

Which intercept is the point at which a graph crosses the y-axis?

If you know two points on a line, what is the slope formula?

What is the slope-intercept form of a linear equation?

What does m represent?   What does b represent?

Caution! The slope m is only the _____.
The _____ is not included in the slope.

Write the standard form of a linear equation:

EXAMPLE 1 Find the slope and y-intercept of $y = \dfrac{2}{3}x + 5$.

Find m.
Find b.

EXAMPLE 2 Find the slope and y-intercept of $y = 4x - 6$.
Find m.
Find b.

EXAMPLE 3 Find the slope and y-intercept of $4x - 3y = 12$.
Rewrite in slope-intercept form.

Find m.

Find b.

EXAMPLE 4 Find the slope and y-intercept of $y = 3$.
Find the slope.

Find the y-intercept.

253

EXAMPLE 5 Find the slope and *y*-intercept of $x = -7$, if possible.
Find the slope.

Find the *y*-intercept.

EXAMPLE 6 Write the equation of a line with a slope of 2 and a *y*-intercept of $\left(0, \dfrac{4}{3}\right)$.

EXAMPLE 7 Write the equation of a line with a slope of –6 and a *y*-intercept of (0, 0).

EXAMPLE 8 Write the equation of the line shown in the graph.

Find the *y*-intercept.
Find the slope.
Substitute.

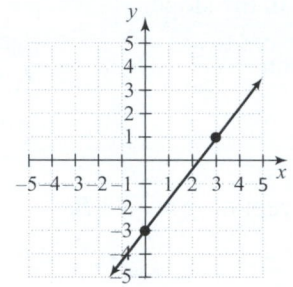

CONCEPT CHECK:
In the equation $y = mx + b$, which letter represents the slope?

GUIDED EXAMPLES:

1. Find the slope and *y*-intercept of the equation $y = \dfrac{6}{7}x + 3$.

2. Find the slope and *y*-intercept of the equation $-5x + y = 3$.

3. Write the equation of a line with a slope $\dfrac{5}{8}$ and a *y*-intercept of (0, 3).

4. Write the equation of a line with a slope of $-\dfrac{2}{3}$ and *y*-intercept of (0, 0).

254

Name: _____   Date: _____

Instructor: _____   Section: _____

Notebook 18.3
Graphing Lines Using the Slope and *y*-Intercept

The _____ is the point where a line crosses the *y*-axis

What are the coordinates of the *y*-intercept?

EXAMPLE 1 Find the *y*-intercept of the graph.

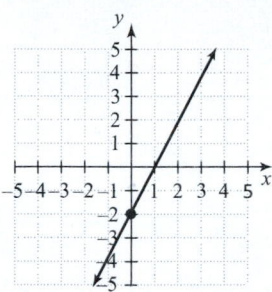

The _____ of a line that passes through two points can be found using the formula $m =$ _____ .

EXAMPLE 2 Find the slope and y-intercept of the line in the graph.

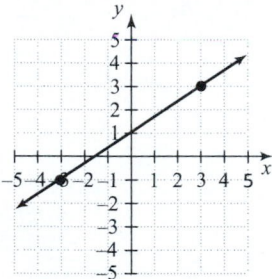

Write the steps for graphing a linear equation with slope and *y*-intercept.
1.

2.

3.

4.

The slope of a line can be described as _____ .

255

EXAMPLE 3 Find the slope and *y*-intercept of the line.

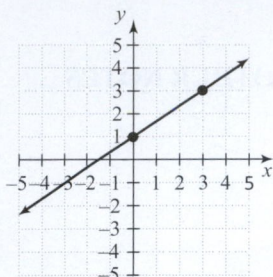

EXAMPLE 4 Graph $y = \frac{1}{2}x - 3$.

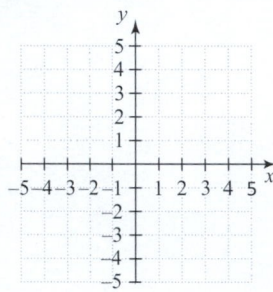

Slope, given as an integer or mixed number,
should be written as _____.
If a slope is 2, write it as _____.

How would you write a slope of $3\frac{5}{8}$?

EXAMPLE 5 Graph $y = -3x$.

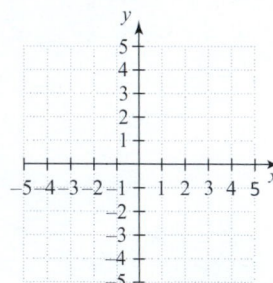

256

EXAMPLE 6 Graph $2x + 3y = 6$.

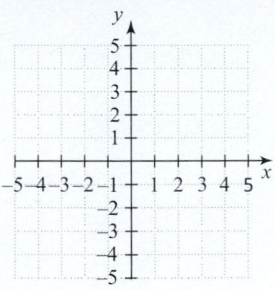

CONCEPT CHECK:

What is best described by $\dfrac{\text{rise}}{\text{run}}$?

GUIDED EXAMPLES:

Graph each equation.

1. $y = \dfrac{3}{4}x + 3$

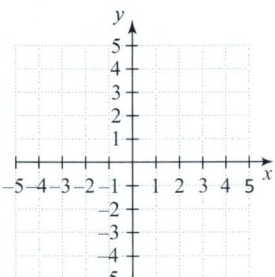

2. $y = -3x + 6$

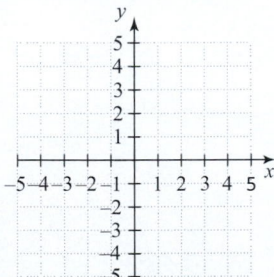

3. $y = x - 2$

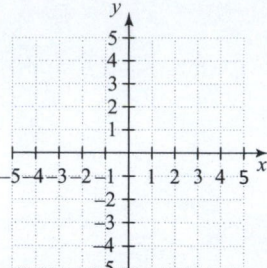

Notebook 18.4
Writing Equations of Lines Using a Point and Slope

Write the slope-intercept form of an equation of a line that has a slope m and a y-intercept of b.

Write the equation of the line with a slope of $-\dfrac{5}{3}$ that passes through the point

$(0, -2)$.

What if the point given is not the y-intercept?

What is the point-slope form of an equation of a line?

What does m represent? What represents the point?

EXAMPLE 1 Use the point-slope form to write the equation of the line with a slope of 2 that passes through $(2, 1)$. Leave your answer in point-slope form.
   Write the point-slope equation.

   Substitute.

EXAMPLE 2 Write the equation of the line with a slope of -3 and passes through $(3, 9)$. Write your answer in slope-intercept form.
   Write the point-slope equation.

   Substitute.

   Solve for y.

EXAMPLE 3 Write the equation of the line which has a slope $\dfrac{1}{7}$ and passes

through $(14, 5)$. Write your answer in slope-intercept form.
   Write the point-slope equation.

   Substitute.

   Simplify.

   Solve for y.

259

EXAMPLE 4 Use the point-slope form to write the equation of the line with a slope of 0 and that passes through (−16, −9). Write your answer in slope-intercept form.

Write the point-slope equation.

Substitute.

Simplify.

Solve for y.

CONCEPT CHECK:

How would $y = 2x - 3$ be written in point-slope form?

GUIDED EXAMPLES:

Write the equation of the line with the given slope and passes through the given point. Write your answer in slope-intercept form.

1. slope of 6 and passes through (−2, −1)

2. slope of $-\dfrac{3}{8}$ and passes through (16, 7)

3. slope of −3 and passes through (−8, −1)

Name: _____   Date: _____

Instructor: _____   Section: _____

Notebook 18.5
Writing Equations of Lines Using Two Points

Write the slope-intercept form of an equation of a line whose slope is m and passes through the point (x_1, y_1).

Examples
Find the slope of the line which passes through (1, 5) and (4, 11).

Write the equation of a line in slope-intercept form with a slope of 2 which passes through (1, 5).

Write the steps for writing the equation of a line given two points.
1.

2.

3.

4.

5.

EXAMPLE 1 Write the equation in slope-intercept form of the line which passes through (2, 5) and (6, 3).
Label the points as (x_1, y_1) and (x_2, y_2).

Find the slope m.

Substitute the values of x_1, y_1, and m into the point-slope form.

Solve for y to write the answer in slope-intercept form.

EXAMPLE 2 Write the equation in of the line in slope-intercept form which passes through (−1, −6) and (4,9).
Label the points as (x_1, y_1) and (x_2, y_2).

Find the slope m.

Substitute the values of x_1, y_1, and m into the point-slope form.

Solve for y to write the answer in slope-intercept form.

261

EXAMPLE 3 Write the equation of the line which passes through $(-5, 5)$ and $(0, -6)$. Write your answer in slope-intercept form.
Label the points as (x_1, y_1) and (x_2, y_2).

Find the slope m.

Substitute the values of x_1, y_1, and m into the point-slope form.

Solve for y to write the answer in slope-intercept form.

EXAMPLE 4 Write the equation of the line which passes through $(-8, 4)$ and $(-5, 0)$. Write your answer in slope-intercept form.
Label the points as (x_1, y_1) and (x_2, y_2).

Find the slope m.

Substitute the values of x_1, y_1, and m into the point-slope form.

Solve for y to write the answer in slope-intercept form.

EXAMPLE 5 Write the equation of the line shown on the graph.
Write your answer in slope-intercept form.

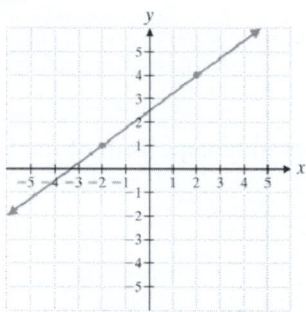

CONCEPT CHECK:
True/False In order to write the equation of a line given two points, one of the points must have an x- or y-coordinate of 0.

GUIDED EXAMPLES:
Write the equation of the line which passes through each pair of points. Write your answer in slope-intercept form.

1. $(6, 5)$ and $(8, 9)$

2. $(3, -5)$ and $(12, 1)$

3. $(2, -3)$ and $(7, -11)$

4. $(-22, -1)$ and $(11, 20)$

262

Name: _____   Date: _____

Instructor: _____   Section: _____

Notebook 18.6
Writing Equations of Parallel and Perpendicular Lines

OTHER NOTES

What are parallel lines?

Parallel lines have _____ slopes but different _____.

What is the symbol for parallel lines?

What is the slope-intercept form of an equation of a line?

What does *m* represent? What represents the *y*-intercept?

EXAMPLE 1 Find the slope of a line parallel to $y = \frac{3}{4}x + 2$.

Find the slope of the line.

Find the slope of a parallel line.

What are perpendicular lines?
Draw lines that are perpendicular.

The slopes of perpendicular lines are _____.
What does that mean?
What is the symbol for perpendicular lines?

EXAMPLE 2 Find the slope of a line perpendicular to $6x + 2y = 9$.
Find the slope of the line.

Find the slope of a perpendicular line.

EXAMPLE 3 Find the slope of a line parallel to the given line and the slope of
a line perpendicular to the given line. $-3x + 5y = -11$
Find the slope of the line.

Find the slope of a parallel line.

Find the slope of a perpendicular line.

EXAMPLE 4 Determine if the lines are parallel, perpendicular, or neither.
$y = 2x + 3$
$-8x + 4y = -4$

EXAMPLE 5 Determine if the lines are parallel, perpendicular, or neither.

$x - 4y = 0$

$y = \dfrac{-1}{3}x - 1$

What is the equation of a horizontal line?
What is the slope of a horizontal line?

What is the equation of a vertical line?
What is the slope of a vertical line?

Write the formula for finding slope of a line containing two points.

EXAMPLE 6 Find the slope of the line containing the points (2, 6) and (4,6). Then find the slope of a line parallel to this line and the slope of a line perpendicular to this line.

Write the point-slope form of an equation.

EXAMPLE 7 Find the equation of a line perpendicular to $y = \dfrac{3}{5}x + 8$ that

passes through the point (3, −1). Write the answer in slope-intercept form.

CONCEPT CHECK:
True/False The slopes of perpendicular lines are negative reciprocals.

GUIDED EXAMPLES:
1. Find the slope of a line parallel and perpendicular to the given line.

   $y = \dfrac{5}{2}x + 16$

2. Determine if the given lines are parallel, perpendicular, or neither.
   $2y = 6x + 8$
   $9x - 3y = -3$

3. Find the slope of a line parallel and perpendicular to $7x - 3y = -18$.

4. Find the equation of a line perpendicular to $4x - 2y = -12$ that passes through the point (−8, 6). Write the answer in slope-intercept form.

Notebook 18.7

Graphing Linear Inequalities in Two Variables

Write an example of a linear inequality.
How many ordered pair solutions are there?

Write the steps for graphing linear inequalities.

1. Replace the _____ with _____.

   _____ the line.

   a. The line will be solid if…

   b. The line will be dotted if …

2. Test the point _____ in the inequality.

   a. If the inequality is true, …

   b. If the inequality is false, …

3. If the point (0, 0) is a point on the line,…

EXAMPLE 1 Graph $4x + 3y \leq 9$.

Graph the line.

Test a point.

Shade.

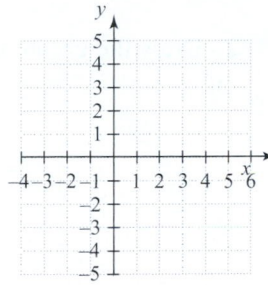

EXAMPLE 2 Graph $3y < -2x$.

Graph the line.

Test a point.

Shade.

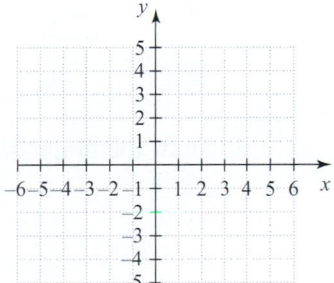

EXAMPLE 3 Graph $y \leq 4$.

Graph the line.

Test a point.

Shade.

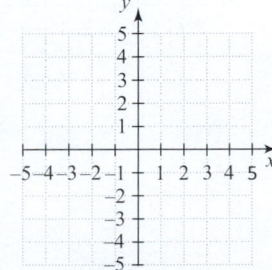

265

EXAMPLE 4 Graph $x > 2$.

Graph the line.

Test a point.

Shade.

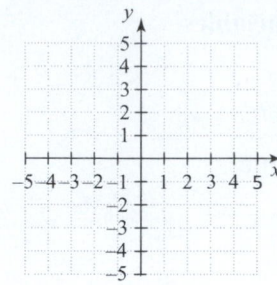

CONCEPT CHECK:

For the inequality $8x < 4y$, is $(2, 1)$ a solution?

GUIDED EXAMPLES:

1. Draw the graph of $2x - y \le 3$.

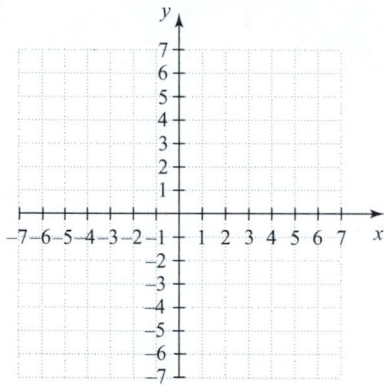

2. Draw the graph of $-4y > 2x$.

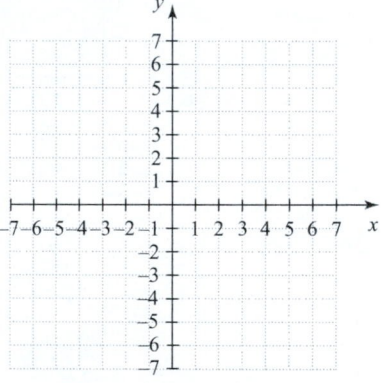

3. Draw the graph of $y \le -2$.

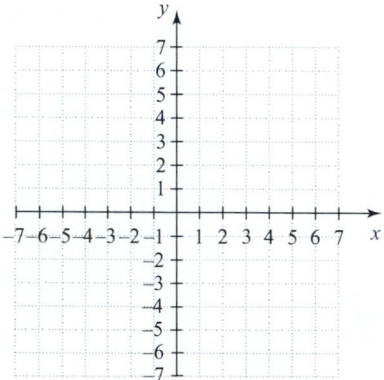

4. Draw the graph of $x > 3$.

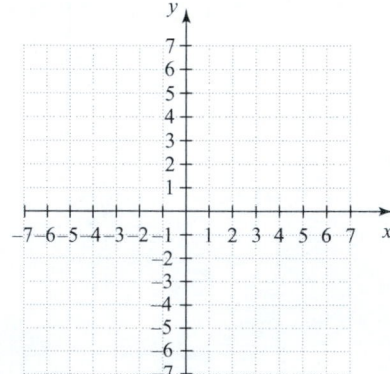

266

Notebook 18.8
Applications of Slope

What is meant by the "rise" of a line?

What is meant by the "run" of a line?

What is the ratio of rise over run?
What variable is used to represent this?

What are some uses of a slope in the real world?

What is the roof pitch?

EXAMPLE 1 Calculate the pitch of this roof.
You will need to sketch the triangle in the roof.

What is the grade (or gradient) used for?

Recall: How do you write a fraction as a percent?

EXAMPLE 2 Write $\frac{3}{8}$ as a percent.

EXAMPLE 3 The Americans with Disabilities Act (ADA) states that the maximum grade for a wheelchair ramp is 5%. The new library will have such a ramp and the first design has the top of the ramp 2 feet higher than the lowest point, with the ramp covering a horizontal distance of 25 feet. Calculate the slope, or grade, of this ramp and determine if it satisfies the ADA requirement.

How do you find the slope of the line passing through two points when you know two points?

267

EXAMPLES 4 Find the slope of the line passing through points (2, −5) and (−8, 6).

EXAMPLE 5 Sky Oaks Summer Camp enroll enrollment during summer 2007 was 491 students. Over the years, the camp became more popular and in 2013 the enrollment had increased to 1235 students. Assuming this increase was linear, calculate the rate at which the enrollment is increasing.

EXAMPLE 6 Using the data shown in the graph below, calculate the slope of the line that represents the manufacturer's suggested retail price (MSRP) for a Toyota Sequoia each year. What does the slope mean with respect to this application?

CONCEPT CHECK:

Is a grade of $\dfrac{1}{10}$ steeper than a grade of $\dfrac{1}{20}$?

GUIDED EXAMPLES:

1. Find the pitch of this roof.

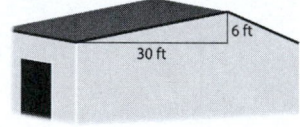

2. It is recommended that a public boat ramp have a grade between 12% and 15% in order to have the best results when launching a boat. The public ramp in Watts Bar Lake covers a horizontal distance of 40 feet and a vertical distance of 6 feet. Does this public boat ramp satisfy the suggested guidelines?

3. Before the first day of school, Alexandira's parents deposited $40 into her lunch account. She purchased the same items for lunch each day for 8 days. At that time, she had $12 left in her account. How much was she spending each day?

268

Notebook 19.1
Relations and Functions

OTHER NOTES

What are three ways that a linear equation in two variables can be represented?

x is called the _____ variable and
y is called the _____ variable.

What is a relation?

What makes up the domain of the relation?
What makes up the range of the relation?

EXAMPLE 1 State the domain and range of the relation.
$\{(5, 7), (9, 11), (10, 7), 12, 14)\}$
domain:

range:

What is a function?

No two _____ ordered pairs can have the same

_____ _____.

EXAMPLES 2 & 3 Determine whether each relation is a function.
$\{(3, 9), (4, 16), (5, 9), (6, 36)\}$

$\{(7, 8), (9, 10), (12, 13), (7, 14)\}$

EXAMPLES 4 – 6 Determine whether each relation is a function.
$y = 3x - 5$

$y = |x|$

$y^2 = x$

269

EXAMPLE 7 Keith makes $10 per hour at his after-school job.
Determine whether the relation between the number of hours he works, x, and his earnings, y, is a function. If it is a function, identify the domain and range.

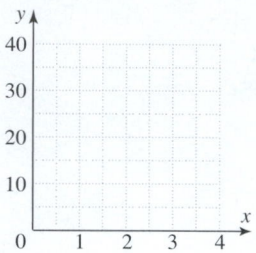

domain:

range:

CONCEPT CHECK: What is a relation for which every x-value in the domain has one and only one y-value?

GUIDED EXAMPLES:
1. State the domain and range of the relation. {(0, 0), (1, 1), (2, 2), (3, 3)}

2. Determine whether the relation is a function. {(6, 7), (7, 8), (8, 7)}

3. Determine whether the relation is a function. $y = |-3x| + 1$

4. Each month, Ashley pays $40 for her cell phone bill plus $0.05 for each text message that she sends. Determine whether the relation between the number of text messages she sends, x, and the amount of her cell phone bill, y, is a function. If it is a function, identify the domain and range.

Name: _____   Date: _____

Instructor: _____   Section: _____

Notebook 19.2
The Vertical Line Test

Can a function have two different ordered pairs with the same first coordinate?
Each value of x must have _____ value of y.

Sketch a graph of a function:

Sketch a graph that does not represent a function:

State the Vertical Line Test (Pencil Test).

How do you use the vertical line test?

EXAMPLE 1 Determine whether the following is the graph of a function.

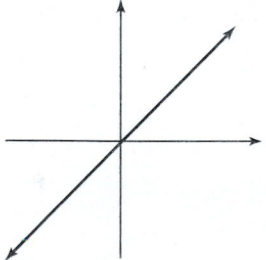

EXAMPLE 2 Determine whether the following is the graph of a function.

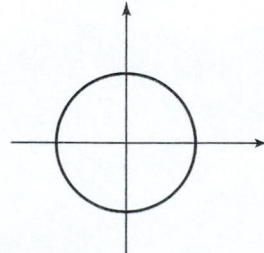

271

EXAMPLE 3 Determine whether the following is the graph of a function.

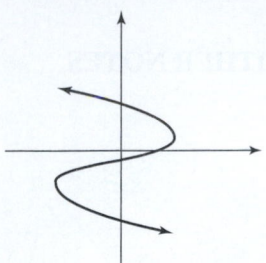

EXAMPLE 4 Determine whether the following is the graph of a function.

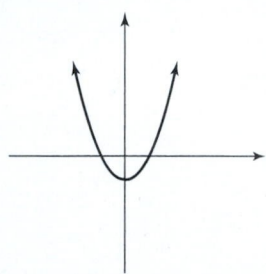

CONCEPT CHECK:

True/False A graph shaped like the letter V would pass the vertical line test.

GUIDED EXAMPLES:
Determine whether each of the following is the graph of a function.

1.
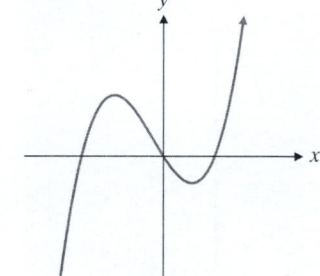

2.
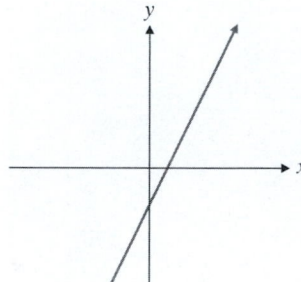

3.
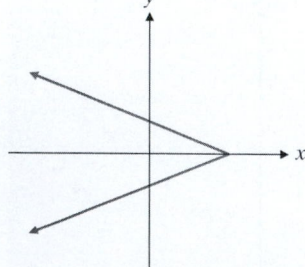

4.
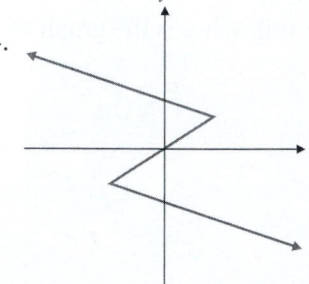

272

Notebook 19.3
Function Notation

If the name of a function is f and the variable is x, what would the function notation be?

How do you read $f(x)$?

EXAMPLES 1 & 2 Use function notation to rewrite the following functions using the given function names.
$y = 9x - 2$, function name f

$y = -16t^2 + 10$, function name h

Note: The variable that is inside the parentheses makes up the _____.
Caution! $f(x)$ does NOT mean _____.

What is the domain of a function?
What is the range?

EXAMPLE 3 Determine the domain and the range of the function.
$f(x) = -4x + 10$

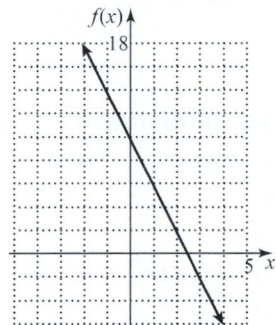

domain:

range:

EXAMPLES 4 & 5 Determine the domain and the range of the function.
$g(x) = x^2 - 4$

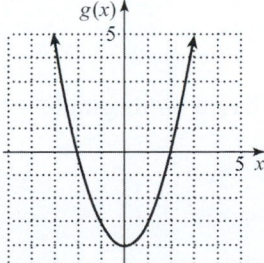

domain:

range:

273

$h(t) = \sqrt{t}.$

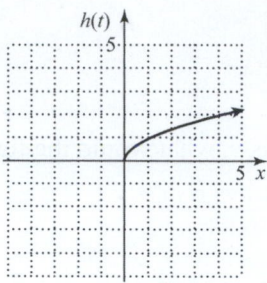

domain:

range:

CONCEPT CHECK:

How can we know that the domain of $f(x) = 3x - 2$ is all real numbers?

GUIDED EXAMPLES:

1. Use function notation to rewrite the function using the given function name.

   $y = 3x + 7$, function name of g

2. Determine the domain and range of the function based on its graph.

   $f(x) = 2x - 5$

   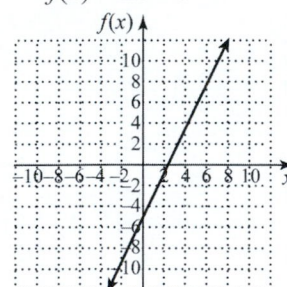

3. Determine the domain and range of the function based on its graph.

   $f(x) = |x|$

   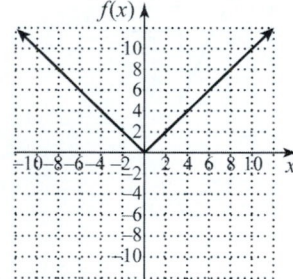

4. Determine the domain and range of the function based on its graph.

   $p(x) = 2x^3 - 6x^2 + 1$

   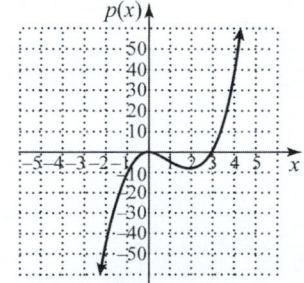

274

Notebook 19.4
Evaluating Functions

What is the procedure for evaluating a function at a certain value?

EXAMPLES 1 & 2 If $f(x) = x + 8$, find the following.

$f(2)$

$f(-6)$

Remember:

$-3^2 =$

$(-3)^2$

$-(-3)^2$

EXAMPLES 3 – 5 If $f(x) = 2x^2 - 4$, find each of the following.

$f(5)$

$f(-3)$

$f(0)$

EXAMPLE 6 The approximate length of a man's femur (thigh bone) is given by the function $f(x) = 0.5x - 17$, where x is the height of the man in inches. Find the approximate length of the femur of a man who is 70 inches tall.

CONCEPT CHECK:

If $f(4) = 6$, what is the ordered pair that is on the graph of the function?

GUIDED EXAMPLES:

1. If $f(x) = 10x - 5$, find $f(-2)$.

2. If $f(x) = x^2 - 2x$, find $f(-2)$.

3. The height of an object dropped from a 100-foot building is given by $h(t) = -16t^2 + 100$, where t is the time after the object is dropped in seconds. Find the height of an object 2 seconds after it is dropped from 100 feet.

Notebook 19.5
Applications of Functions

A relation for which every x-value in the domain has one and only one y-value is called a _____.

If the name of the function is f and the variable is x, how can this function be represented?

Caution! $f(x)$ does not mean _____.

EXAMPLE 1 A turbine wind generator produces P kilowatts of power for a wind speed w, according to the equation $P = 2.5w^2$. Write the number of kilowatts P as a function of the wind speed w, and find the number of kilowatts generated when the wind speed is 12 mph.

EXAMPLE 2 The surface area S of a sphere is given by the equation $S = 4\pi r^2$, where r is the radius of the sphere. Write the surface area as a function of the radius, and find the surface area of a sphere with a radius of 11 cm. Use 3.14 for π.

EXAMPLE 3 A cell phone company has a text-messaging plan that charges $6.50 per month, plus $0.04 for each text message sent. Write C, the cost per month, as a function of t, the number of text messages sent.

What would the monthly cell phone bill be if you sent 2386 text messages in a month?

How many text messages were sent if the bill was $149.98?

EXAMPLE 4 The number of ticket sales to U.S. movie theaters (in billions of tickets) can be estimated by the linear function $f(t) = -0.04t + 1.62$, where t is the number of years after 2002. Find the estimated number of tickets sold in 2011.

EXAMPLE 5 The number of ticket sales to U.S. movie theaters (in billions of tickets) can be estimated by the linear function $f(t) = -0.04t + 1.62$, where t is the number of years after 2002. Predict the year in which the number of tickets sold will be less than 1 billion.

What are the steps for graphing a linear function given the slope and the y-intercept?

EXAMPLE 6 When buying hamburger meat for a cookout, Jerry decides that he will need $\frac{1}{3}$ pound of hamburger per person, and he will buy one extra pound of meat to ensure he has enough. This can be represented as the function $f(x) = \frac{1}{3}x + 1,$ where x is the number of people coming to the cookout.

Graph this function.

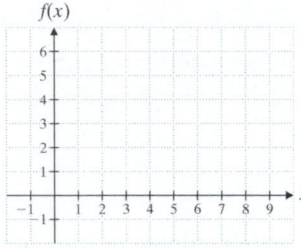

Find the amount of meat Jerry will need to buy if 9 people come to the cookout.

CONCEPT CHECK: A taxi company charges an initial fee of $5.00 and a rate per mile of $1.20. What is the equation that will find the cost of a 10-mile taxi ride?

GUIDED EXAMPLES:
1. The volume V of a cube is given by the equation $V = s3$, where s is the length of a side. Write the volume of a cube as a function of the side length. Find the volume of a cube with a side length of 7 inches.

2. A taxi company charges an initial fee of $4.25 and a rate per mile of $0.75. Write C, the cost for a taxi ride as a function of m, the number of miles ridden. Find the cost of a 13.2-mile taxi ride.

3. A taxi company charges an initial fee of $4.25 and a rate per mile of $0.75. The total amount charged for a ride can be found with the function $C(m) = 0.75m + 4.25.$ If the charge for a ride was $8.90, find the length of the taxi ride.

Notebook 20.1
Introduction to Systems of Linear Equations

In order to be a solution to a system, the ordered pair must make each equation in the system _____.

When observing the graphs of these equations, where is the solution?

What is a system of linear equations?

What is a solution to a system of equations?

Every solution to a system of equations has two parts, which are written together as an ordered pair. What are the two parts?

EXAMPLE 1 Determine whether $(3, -2)$ is a solution to the following system of equations.
$$x + 3y = -3$$
$$4x + 3y = 6$$

EXAMPLE 2 Determine whether $(4, 3)$ is a solution to the following system of equations.
$$7x - 4y = 16$$
$$5x + 2y = 24$$

EXAMPLE 3 Determine whether $(-9, -7)$ is a solution to the following system of equations.
$$3y = 2x - 3$$
$$y = x - 5$$

EXAMPLE 4 Determine whether $\left(\dfrac{4}{3}, \dfrac{1}{6}\right)$ is a solution to the following system of equations.

$$2y = 1 - \frac{1}{2}x$$
$$3x = 2 + 12y$$

EXAMPLE 5 Determine whether $(7, 6)$ is a solution to the following system of equations.

$$0.3x + 0.2y = 2.7$$
$$9x - 3y = 4$$

CONCEPT CHECK:

How do you check the solution for a system of equations?

GUIDED EXAMPLES:

1. Determine whether $(-6, -11)$ is a solution to the system.
$$3x - 2y = 4$$
$$-9x + 3y = 21$$

2. Determine whether $(5, 1)$ is a solution to the system.
$$x + 8y = 12$$
$$-3x - y = -16$$

3. Determine whether $(13, -2)$ is a solution to the system.
$$0.2x + 1.5y = -0.4$$
$$3x - 2y = 35$$

4. Determine whether $\left(-\dfrac{2}{3}, -\dfrac{3}{5}\right)$ is a solution to the following system of equations.
$$6x = 5y - 1$$
$$\frac{5}{3}y = -2x + 1$$

Name: _____   Date: _____

Instructor: _____   Section: _____

Notebook 20.2
Solving by the Graphing Method

OTHER NOTES

The solution to a system of linear equations is the
_____ of the lines.

How do you check that your solution is correct?

EXAMPLE 1 Find the solution (the point of intersection of the two lines).
Always check your solution.

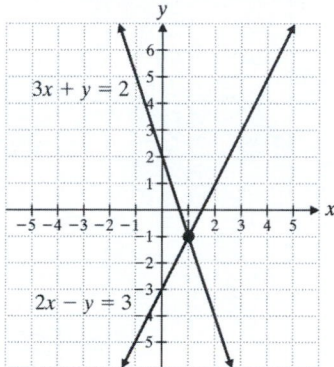

Write the steps for solving a system using the Graphing Method.

1.

2.

3.

4.

If the lines intersect, what is the solution?

If the lines do not intersect, what is the solution?

If the equations are the same line, what is the solution?

Sketch the graphs of these three cases:

EXAMPLE 2 Find the solution to the system of equations by graphing.

$$3x + 2y = 10$$
$$x - y = 5$$

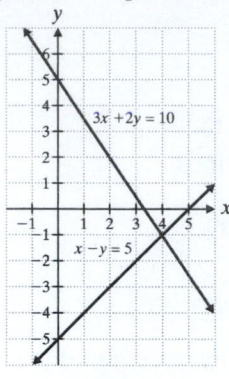

EXAMPLE 3 Find the solution to the system of equations by graphing.

$$y = x - 1$$
$$2x + y = 8$$

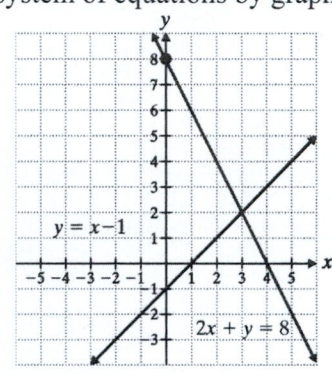

What is an inconsistent system of equations?
Give an example of an inconsistent system.

What is a dependent system of equations?
Give an example of a dependent system.

CONCEPT CHECK:

What kind of system has no solution? Its graph will be two parallel lines that do not intersect.

GUIDED EXAMPLES:

1. Solve using the graphing method.

$$3x - y = 5$$
$$2x - 3y = -6$$

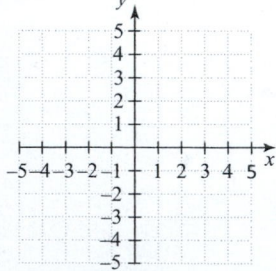

2. Solve using the graphing method.

$$3x - 2y = 6$$
$$4x + y = -3$$

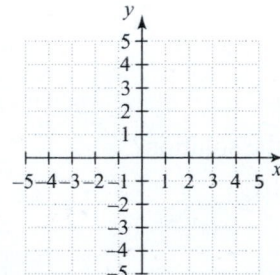

282

Name: _____   Date: _____

Instructor: _____   Section: _____

Notebook 20.3
Solving by the Substitution Method

OTHER NOTES

When solving a system of equations using the graphing method, the solution is the _____.

This could be difficult if the solution is a _____.

Thus, the need for the substitution method.
Write the steps for solving systems using substitution.

1. Choose _____ equation and _____ for either variable in terms of the other. _____ this equation.

2. _____ the expression from step 1 into the _____ equations.

3. _____ this equation.

4. _____ this value in place of the _____ into either of the _____ equations to obtain a value for the second _____. _____ this equation.

5. Write the values for both variables in an _____.

6. _____ the solution in _____ equations.
Note: What happens if, after step 3, you get an equation with no variables?
A true result means …

A false result means…

EXAMPLE 1 Solve the system of equations by substitution. Then check.
$$4x + 3y = 50$$
$$y = 2x$$

EXAMPLE 2 Solve the system of equations by substitution. Then check.
$$y = x - 5$$
$$3x - 2y = 17$$

EXAMPLE 3 Solve the system of equations by substitution. Then check.
$$-4x + 3y = 10$$
$$-6x + y = 1$$

Note: Again, what happens if you get an equation with no variables and the result is true?
the result is false?

EXAMPLE 4 Solve the system of equations by substitution. Then check.
$$4x - 24y = 40$$
$$x - 6 = 10$$

EXAMPLE 5 Solve the system of equations by substitution. Then check.
$$x + 3y = 5$$
$$4x + 12y = 40$$

CONCEPT CHECK:
After a system of equations has been solved, what would a result of $2 = 0$ mean?

GUIDED EXAMPLES:
Solve the system of equations by substitution. Then check.

1. $3x + 2y = 15$
   $y = 6x$

2. $2x + y = -1$
   $3x + 3y = 15$

3. $y = 3x + 1$
   $6x - 2y = -2$

4. $y = -2x + 3$
   $2x + y = 6$

Notebook 20.4
Solving by the Elimination Method

A _____ is a set of two or more equations with the same variables.

The _____ to a system of equations is the _____ that is a solution to every equation in the system.

The solution to a system of equations has 2 parts, which are written together as an _____.

What are opposite numbers?

EXAMPLE 1 Solve the system of equations. Then check.

$$7x - 3y = 1$$
$$5x + 3y = 11$$

Write the steps for solving systems by elimination.
1. Arrange each equation in the form _____.

2. If necessary, _____ one or both _____ by the appropriate number so that the _____ of either variable are _____.

3. _____ the two equations so that the variable is eliminated.

4. _____ the resulting _____.

5. _____ this value into either of the _____ _____ and _____ to find the value of the other variable.

6. _____ the solution as an _____.

7. _____ the solution in both of the _____.

Note: What happens if, after step 3, you get an equation with no variables?
A true result means …
A false result means…

EXAMPLE 2 Solve the system of equations by elimination. Then check.

$$3x + 7y = 22$$
$$3x - 2y = 3$$

EXAMPLE 3 Solve the system of equations by elimination. Then check.

$$4x - 10y = -44$$
$$9x - 5y = 6$$

Note: An easy way to know what numbers to multiply each equation is to:
_____ which _____ you want to _____;
_____ each equation by the _____ of that variable term
in the _____ equation.

Caution! Make sure you multiply _____ sides of the equation by the number.

EXAMPLE 4 Solve the system of equations by elimination. Then check.

$$4x + 3y = -5$$
$$7x + 2y = 14$$

EXAMPLE 5 Solve the system of equations by elimination. Then check.

$$3x + 2y = 18$$
$$-3x - 2y = 14$$

EXAMPLE 6 Solve the system of equations by elimination. Then check.

$$9x - 6y = 24$$
$$3x - 2y = 8$$

CONCEPT CHECK:
What is the first step in solving a system of equations using elimination?

GUIDED EXAMPLES:
Solve the system of equations by elimination. Then check.

1. $8x + 4y = -4$
   $-8x - 7y = -11$

2. $5x + 4y = 96$
   $9x - 3y = 81$

3. $6x + 9y = 24$
   $4x + 6y = 15$

4. $12x - 5y = -7$
   $4x + 2y = 5$

Notebook 20.5
Solving a System in Three Variables by the Elimination Method

A _____ is a set of two or more equations containing the same variables.

What type of solution does a system of linear equations in two variables have?

If the system of equations has three variables, then what kind of solution will that system have?

Write the steps for solving a system in three variables by elimination.
1. Choose _____ and use _____ to eliminate one of the variables.

2. Choose a _____ set of _____ and use the _____ to eliminate the _____ variable as step 1.

3. _____ the resulting system of _____ equations in _____.

4. _____ the _____ from step 3 into one of _____ equations and _____ for the remaining variable.

5. _____ your solution in ____ of the original equations. In order to be correct, it must satisfy _____ the equations.

EXAMPLE 1 Solve the system of equations using elimination. Then check.
$$-2x + 5y + z = 8$$
$$x + 3y - z = 5$$
$$-x + 2y + 3z = 13$$

EXAMPLE 2 Solve the system of equations using elimination. Then check.
$$4x + 3y + 3z = 4$$
$$3x \qquad + 2z = 5$$
$$2x - 5y \qquad = 13$$

What are the three possibilities for a solution to a system of equations
in three variables?

If the three planes intersect at one point, then…

If the three planes intersect at a common line or if they are the
same plane, then…

If the planes do not all intersect, then…

EXAMPLE 3 Solve the system of equations using elimination. Then check.
$$6x - 2y + 2z = 2$$
$$4x + 8y - 2z = 5$$
$$-4x - 8y + 2z = -4$$

EXAMPLE 4 Solve the system of equations using elimination. Then check.
$$-28x - 14y \quad\quad = 42$$
$$\quad\quad 2y + 16z = 5$$
$$14x + 6y - 8z = -12$$

CONCEPT CHECK:
Bill solves a system of equations in three variables and in one step, receives an
answer of $0 = 0$. How should he interpret this solution?

GUIDED EXAMPLES:
1. Determine whether $(1, -8, 5)$ is a solution to the system.
$$2x + y - 3z = -21$$
$$-4x - 2y + 7z = 47$$
$$3x + 6y - 4z = -65$$

2. Solve.
$$2x + y + 3z = 2$$
$$x - y + 2z = -4$$
$$x + 3y - z = 1$$

3. Solve.
$$3x + 3y - 3z = -1$$
$$4x + y - 2z = 1$$
$$-2x + 4y - 2z = -8$$

288

Name: _____ Date: _____

Instructor: _____ Section: _____

Notebook 20.6
Applications of Systems of Linear Equations

Write the steps for solving applications with a system of equations:
1.

2.

3.

4.

5.

6.

EXAMPLE 1 A movie theater sells tickets for $10 and bags of popcorn for $3. If a single Saturday night, the theater had $2375 in sales. The theater owner found that if he raised ticket prices to $11, raised popcorn prices to $4, and sold the same number of tickets and popcorn, the theater would make $2700. How many tickets were sold? How many bags of popcorn were sold?
Define the variables.
Write a system of equations.
Solve.

EXAMPLE 2 A boat travels 20 miles upstream, against the current, in 4 hours. The return trip, 20 miles downstream, with the current, only takes 2 hours.
Find the speed of the boat in still water and the speed of the current.
Let r = the speed of the boat in still water, and c = the speed of the river current.
Note: Distance = _____ × _____

	Distance	Rate	Time	Equation
Upstream				
Downstream				

EXAMPLE 3 An electronics company makes two types of switches. Type A takes 4 minutes to make and requires $3 worth of materials. Type B takes 5 minutes to make and requires $5 worth of materials. In the latest production batch, it took 35 hours to make these switches, and the materials cost $1900. How many of each type of switch were made?

EXAMPLE 4 A trucking company has three sizes of trucks. A large truck holds 10 tons of gravel, a medium-sized truck holds 6 tons, and a small truck holds 4 tons. The company wants to use fifteen trucks to haul 104 tons of gravel. To save money on gas, the company's manager wants to use two more large trucks than medium-sized trucks. How many of each type of truck will be used?

CONCEPT CHECK:
What is the first step in solving applications involving systems of linear equations?

GUIDED EXAMPLES:

1. Concert tickets bought in advance cost $5, but at the door they are $6. The ticket sales came to $4540. The concert promoter raises tickets for the next show to $7 in advance and $9 at the door. If the same number of people attend the next concert, the ticket sales will total $6560. Find how many tickets were sold in advance and how many were bought at the door.

2. A boat travels 105 km downstream in 5 hours. The return trip upstream takes 7 hours. Find the speed of the boat in still water and the speed of the current.

3. A furniture company makes large and small chairs. A small chair takes 30 minutes of machine time and 75 minutes of labor to build. A large chair takes 40 minutes of machine time and 80 minutes of labor to build. The company has 57 hours of labor time and 26 hours of machine time available each day. How many of each type of chair is built in a day?

Name: _____   Date: _____

Instructor: _____   Section: _____

Notebook 21.1
Introduction to Polynomials

A _____ is any number, variable, or product of numbers and/or variables.
A _____ is a number that is multiplied by a variable.
Terms that have the same variable(s) raised to the same exponent(s) are called

_____.

Reminder: How do you combine like terms?

What is a polynomial in the variable x?

What does it mean to write a polynomial in descending order?

What is the degree of a term?

What if there is more than one variable, then how do you get the degree of the term?

The degree of any constant (numerical) term is _____.

EXAMPLES 1–3 Find the degree of each term.
$7x^3$

$4ab^2$

10

What is the degree of a polynomial?

EXAMPLES 4 & 5 For each polynomial, find the degree of each term.
Then find the degree of the polynomial.
$5x^3 + 8x^2 - 20x - 2$

$6xy - 4x^2y + 2xy^3$

There are special names for polynomials with _____, _____,
or _____ terms.

A _____ is a polynomial with exactly _____ term. Example:

A _____ is a polynomial with exactly _____ terms. Example:

A _____ is a polynomial with exactly _____ terms. Example:

OTHER NOTES

291

EXAMPLES 6–8 State the degree of the polynomial, and state whether each polynomial is a monomial, a binomial, or a trinomial.

$5x + 3x^3$

$-7a^5b^2$

$8x^4 - 9x - 15$

What is the procedure for evaluating a polynomial?

EXAMPLE 9 Evaluate the polynomial $P(x) = 2x^2 - 6x + 7$ at $x = -2$.

EXAMPLE 10 The height of an object dropped from a 200 ft building is given by the polynomial $h(t) = -16t^2 + 200$, where t is the time after the object is dropped in seconds. Find the height of an object 3 seconds after it is dropped from 200 feet.

CONCEPT CHECK:
True/False The polynomial $5x^4 + 6x$ has a degree of 4.

GUIDED EXAMPLES:
1. Find the degree of each term. $3x^4$ and $2xy^2$

2. Find the degree of the polynomial. $7xy^2 - 2x^2y^2 + 5y$

3. Evaluate the polynomial. $P(x) = -4x^3 + 2x^2 - x + 5$

4. The concentration C, in parts per million, of a certain antibiotic in the bloodstream after t hours is given by the polynomial $C(t) = -0.05t^2 + 2t + 2$. Find the concentration after 4 hours.

Notebook 21.2
Addition of Polynomials

Terms that have the same variable(s) raised to the same exponent(s) are called

_____.

Like terms can be combined by _____ or _____
the coefficients of the terms. Example.

What is the rule for adding polynomials?

EXAMPLE 1 Add. $(x + y) + (2x + 6y)$

EXAMPLE 2 Add. $(5x^2 - 6x - 2) + (-3x^2 - 9x + 5)$

EXAMPLE 3 Add. $(7x^2 + 8x + 9) + (13x^2 - 10x + 5)$

EXAMPLE 4 Add. $(1.2x^3 - 5.6x^2 + 5) + (-3.4x^3 - 1.2x^2 + 4.5x - 7)$

EXAMPLE 5 Add. $\left(\dfrac{1}{2}x^2 - 6x + \dfrac{1}{3}\right) + \left(2x - \dfrac{1}{2} + \dfrac{1}{5}x^2\right)$

EXAMPLE 6 Add. $(6x^2 - x + 4) + (10x - 8x^2 - 3)$

CONCEPT CHECK:
What is the simplest form of $5.1x^2 - 2x^2 + 3x + 0.5x - 12 + 5$?

GUIDED EXAMPLES:
Add each of the following.

1. $(-8x^3 + 3x^2 + 6) + (2x^3 - 7x^2 - 3)$

2. $(12x^2 - 8x - 11) + (-7x^2 + 6x - 13)$

3. $(3.5x^3 - 0.02x^2 + 1.56x - 3.5) + (-0.08x^2 - 1.98x + 4)$

4. $\left(-\dfrac{1}{3}x^2 - 6x - \dfrac{1}{12}\right) + \left(5x + \dfrac{1}{4}x^2 - \dfrac{1}{3}\right)$

Notebook 21.3
Subtraction of Polynomials

To add two polynomials, _____.

Subtraction can be defined as _____.

What is the procedure for subtracting polynomials?

EXAMPLE 1 Subtract. $(2x + 3) - (x - 5)$
Change the sign of each term in the second polynomial.

Add by combining like terms.

EXAMPLE 2 Subtract. $(-2x^3 + 7x^2 - 3x - 1) - (-6x^3 - 9x^2 - x + 4)$

EXAMPLE 3 Subtract. $(-3x^4 + 5x^2 + 2) - (6x^3 - 10x^2 + 2x - 1)$

EXAMPLE 4 Subtract. $(0.7x^3 - 1.2x - 4.8) - (1.6x^3 + 9.4x - 6.5)$

Caution! Use extra care in determining which terms are _____
when polynomials contain more than one variable.

EXAMPLE 5 Subtract. $(-6x^2y - 3xy + 7xy^2) - (5x^2y - 8xy - 15x^2y^2)$

CONCEPT CHECK:

What is the first step when subtracting polynomials?

GUIDED EXAMPLES:

Subtract each of the following.

1. $(-8x^2 + 3x - 7) - (5x^2 - 6x + 2)$

2. $(12x^5 - 3x^3 + 8x - 6) - (-4x^5 + 13x^4 + 5x^2 - 10x + 4)$

3. $(7.2x^2 - 1.4x - 4.5) - (2.3x - 0.8x^2 + 3.6)$

4. $(a^3 - 7a^2b + 3ab^2 - 2b^3) - (2a^3 + 4ab - 6b^3)$

Name: _____   Date: _____

Instructor: _____   Section: _____

Notebook 21.4
Product Rule for Exponents

OTHER NOTES

An _____ is used as a shortcut for repeated multiplication.
The number being multiplied is the _____.
The number of times the base is used as a factor is the _____.

What is an exponential expression?

What happens when we multiply exponential expressions? Give examples.

What happens to the base?
What happens to the exponent?

What does the product rule for exponents say about multiplying exponential expressions that have like bases?

EXAMPLES 1–3 Multiply.
$x^3 \cdot x^6$

$y \cdot y^5$

$x^3 \cdot x \cdot x^6$

Caution! What do you do if the variable does not have a written exponent?

Caution! The product rule for exponents applies only to expressions that
_____.

EXAMPLES 4–6 Simplify, if possible.
Write your answer using exponential notation..
$y^5 \cdot y^{11}$

$2^3 \cdot 2^5$

$x^6 \cdot y^8$

How do you multiply exponential expression with coefficients?

297

EXAMPLES 7–9 Multiply.
Write your answer using exponential notation.
$(3a)^2(3a)^4$

$(5x^3)(x^6)$

$(-6x)(-4x^5)$

What do you do in cases where the problems involve more than one variable or more than two factors?

EXAMPLE 10 Multiply. $(5ab)\left(-\dfrac{1}{3}a\right)(9b^2)$

CONCEPT CHECK:
The product rule can be applied when _____ exponential expressions with like bases. (Fill in which multiplying, dividing or adding.)

GUIDED EXAMPLES:
Subtract each of the following.
1. Multiply. $w^{10} \cdot w$

2. Simplify by the product rule, if possible. $(4x)^3(4x)^7$

3. Multiply. $(-4x^3)(-5x^2)$

4. Multiply. $(2xy)\left(-\dfrac{1}{4}x^2y^2\right)(6xy^3)$

Name: _____   Date: _____

Instructor: _____   Section: _____

Notebook 21.5
Power Rule for Exponents

OTHER NOTES

An exponent is used as a shortcut for _____.
The base is the number being _____.
The exponent is the number of times the base is used as a _____.

An _____ is a variable or a number raised to
an exponent.

To multiply two exponential expressions that have _____,
_____ the base and _____ the exponents.
Example:

What does the power rule for exponents say about raising a product to a power?

EXAMPLES 1–3 Simplify. Write your answer using exponential notation.

$(x^3)^5$

$(2^7)^3$

$(y^2)^4$

How do you raise a product to a power?
Example:

EXAMPLES 4–6 Simplify.

$(ab)^8$

$(3x)^4$

$(-2x^2)^3$

How do you raise a fractional expression (quotient) to a power?
Example:

EXAMPLES 7 & 8 Simplify. Write your answer using exponential notation.

$$\left(\frac{x}{y}\right)^5$$

$$\left(\frac{3}{w}\right)^4$$

Caution! As long as you use the rules _____, you can apply them in _____ order. Be sure to take care to determine the correct _____ of the expression, especially if there is a _____.

EXAMPLE 9 Simplify. $\left(\dfrac{-3x^2z^0}{y^3}\right)^4$

CONCEPT CHECK:

How would the z variable in $(-3x^2yz^0)^3$ be simplified?

GUIDED EXAMPLES:

1. Simplify. Write your answer using exponential notation. $(y^3)^7$

2. Simplify. $(2y^4)^2$

3. Simplify. $\left(-\dfrac{7a}{b}\right)^2$

4. Simplify. $\left(\dfrac{2y^0z^2}{4x^3}\right)^5$

Name: _____   Date: _____

Instructor: _____   Section: _____

Notebook 22.1
Multiplying by a Monomial

For all real numbers a, b, and c, $a(b + c) =$ _____.
To multiply two exponential expressions that have _____,
_____ the base and _____ the exponents.
In symbols, this looks like …

What does the power rule for exponents say about raising a product to a power?

Give some examples of monomials.

A _____ is a polynomial with exactly one term.

What is the rule for multiplying a monomial by a polynomial?

EXAMPLE 1 Multiply. $3x^2(5x - 2)$

EXAMPLE 2 Multiply. $2x(x^2 + 3x - 1)$

EXAMPLE 3 Multiply. $-6xy(x^3 + 2x^2y - y^2)$

EXAMPLE 4 Multiply. $3a(2a^3 - 3a^2 + 7a)$

301

EXAMPLE 5 Multiply. $(2x^2 - 3x + 8)(-7x)$
Rewrite using Commutative Property.

Note: Why did we change the order here?

CONCEPT CHECK:
Use the Distributive Property to simplify $3(x - 4)$.

GUIDED EXAMPLES:
1. Multiply. $4x^3(-2x^2 + 3x)$

2. Multiply. $3x(x^2 + 2x - 4)$

3. Multiply. $-2xy^2(x^2 - 5xy - 3y^2)$

4. Multiply. $(x^3 - 2x + 6)(-2x)$

Notebook 22.2
Multiplying Binomials

A _____ is a polynomial with exactly two terms.
Give some examples of binomials.

Write the steps for multiplying binomials.
1.

2.

3.

EXAMPLE 1 Multiply. $(3x + 1)(x + 4)$

EXAMPLE 2 Multiply. $(7x - 4)(6x + 3)$

What happens if one or more of the binomials involve subtraction?

EXAMPLE 3 Multiply. $(4x - 9y)(8x - 3)$

Does the process of multiplication change if there is more than one variable in a binomial?

How is the FOIL method used in multiplying binomials?
F:
O:
I:
L:

303

Can the FOIL method be used to multiply any binomials?
If not, which ones?

EXAMPLE 4 Multiply using the FOIL method. $(x + 3)(x + 5)$
First:
Outer:
Inner:
Last:

EXAMPLE 5 Multiply using the FOIL method. $(2x + 1)(3x + 4)$
First:
Outer:
Inner:
Last:

EXAMPLE 6 Multiply using the FOIL method. $(-2 + 7x)(-9x + 5)$
First:
Outer:
Inner:
Last:

CONCEPT CHECK:
Which of the following is a binomial?. $x^2 + 7$ or $3a^2 + 4a - 5$

GUIDED EXAMPLES:
1. Multiply. $(x + 2)(5x + 8)$

2. Multiply. $(2x - 15y)(7x - 2)$

3. Multiply using the FOIL method. $(x^2 + 2)(x^2 + 9)$

4. Multiply using the FOIL method. $(3x^2 - 2)(-4x^2 + 5)$

Name: _____   Date: _____

Instructor: _____   Section: _____

Notebook 22.3
Multiplying Polynomials

The _____ can be used to multiply a polynomial by a monomial.

Ex. Multiply. $3x(4x^2 + x - 2)$

The Distributive Property can also be used to multiply two _____.

Ex. Multiply. $(2x + 3)(4x + 5)$

What does it mean if a polynomial in x is written in descending order?

Write the steps for multiplying polynomials.

1.

2.

3.

4.

EXAMPLE 1 Multiply. $(x + 4)(3x^2 + x + 2)$
Multiply each term in the first
polynomial by every term in the
second polynomial.

Write the sum.
Combine like terms.
Descending order.

EXAMPLE 2 Multiply. $(3x - 1)(6x^2 - 5x + 8)$

EXAMPLE 3 Multiply. $(5x^5 + 2x^2 + 7)(x^4 + x)$

EXAMPLE 4 Multiply. $(4x^2 + 9x + 7)(2x^2 - 6x - 5)$

Integers, as well as polynomials, can be multiplied _____.
Multiply $(x^2 + 3x + 8)(2x^2 + 4x - 5)$ vertically.

How do you multiply three or more polynomials?

EXAMPLE 5 Multiply. $(x + 1)(x + 2)(x + 3)$

CONCEPT CHECK: Is $3x^2 + 9x^3 - 2x^4 + 7x^5$ written in descending order?

GUIDED EXAMPLES:

1. Multiply. $(9x + 5)(x^2 + 7x - 3)$

2. Multiply. $(3x^2 - 4x - 6)(2x - 3)$

3. Multiply. $(x^2 + 3x + 4)(2x^2 - x - 7)$

4. Multiply. $(3x + 2)(5x - 7)(x + 1)$

Name: _____   Date: _____

Instructor: _____   Section: _____

Notebook 22.4
Multiplying the Sum and Difference of Two Terms

OTHER NOTES

To multiply two binomials, multiply each term in the _____ binomial by each term in the _____ binomial, or use _____.

Multiply $(x + 3)(x - 3)$.

EXAMPLE 1 Multiply. $(5x + 4)(5x - 4)$

What are some observations from the pattern?

What is the rule for multiplying binomials for a sum and a difference?

EXAMPLES 2 & 3 Multiply.
$(x + 5)(x - 5)$

$(7a - 8)(7a + 8)$

EXAMPLES 4 & 5 Multiply.
$(2x^2 + 3y)(2x^2 - 3y)$

$\left(\dfrac{1}{4}x - \dfrac{2}{3}\right)\left(\dfrac{1}{4}x + \dfrac{2}{3}\right)$

Show why this formula for multiplying a sum and difference works.

Use formula (shortcut) $(a - b)(a + b) =$

Use FOIL $(a - b)(a + b) =$

CONCEPT CHECK: Write the formula for multiplying the sum and difference of two terms.

GUIDED EXAMPLES:

1. Multiply. $(x - 2)(x + 2)$

2. Multiply. $(6x + 1)(6x - 1)$

3. Multiply. $(7x^3 + 12y)(7x^3 - 12y)$

4. Multiply. $(xy - 7)(xy + 7)$

Notebook 22.5
Squaring Binomials

To multiply two binomials, multiply each term in the _____ binomial by each term in the _____ binomial, or use _____.

Multiply $(x + 7)(x + 7)$.

Caution! $(x + 7)^2 = (x + 7)(x + 7)$
$(x + 7)^2$ does not mean $x^2 + 7^2$

EXAMPLE 1 Simplify.
$(2x - 3)^2$

Write as a multiplication problem.
Multiply the binomials.
Simplify.

What are some observations from the pattern?

What are the rules for squaring a binomial?
$(a + b)^2 =$
$(a - b)^2 =$

EXAMPLE 2 Simplify.
$(3x + 5)^2$

EXAMPLES 3 & 4 Simplify.

$(x - 5)^2$

$(2x^2 + 3y)^2$

EXAMPLE 5 Simplify.

$\left(\dfrac{1}{4}x - \dfrac{2}{3}\right)^2$

CONCEPT CHECK: Write the formula for squaring a binomial.

GUIDED EXAMPLES:

1. Multiply. $(3x + 11)^2$

2. Multiply. $(8x^3 + 9)^2$

3. Multiply. $(6x - 1)^2$

4. Multiply. $\left(\dfrac{3}{8}y - \dfrac{1}{10}\right)^2$

Name: _____   Date: _____

Instructor: _____   Section: _____

Notebook 23.1
The Quotient Rule

To simplify a fraction using the prime factors method,
1. Write the _____ and _____ as the
   product of _____-.
2. Divide by _____ in the numerator and denominator.
3. Multiply the remaining _____ to get simplest form.

Example. Simplify using the prime factors method.
$$\frac{64}{16}$$

Why do you need to review simplifying fractions?

EXAMPLE 1 Simplify using the prime factors method.
$$\frac{x^6}{x^4}$$

EXAMPLE 2 Simplify using the prime factors method.
$$\frac{25x^6}{10x^3}$$

What does the Quotient Rule say?
To _____ like bases, _____ the base and _____
the exponents. In symbols, …

EXAMPLES 3–5 Simplify.
$$\frac{x^3}{x^2}$$

$$\frac{42y^5}{18y^3}$$

$$-\frac{16s^6}{32s^2}$$

EXAMPLE 6 Simplify using the quotient rule.
$$\frac{7^{10}}{7^4}$$

Caution! The _____ does not change, only the _____ changes.

311

EXAMPLE 7 Simplify using the quotient rule.

$$\frac{y^5}{y^5}$$

Remember, any non-zero number divided by itself is _____.

What does the "Zero as an Exponent Property" say?

Note: What about 0^0?

EXAMPLES 8–10 Evaluate. Assume all variables represent nonnegative numbers.

$$\frac{8^5}{8^5}$$

$$\frac{(ax)^4}{(ax)^4}$$

$$\frac{5x^3}{x^3}$$

CONCEPT CHECK: Apply the quotient rule to $\dfrac{5x^4 y^2}{x^2 y}$.

GUIDED EXAMPLES:

1. Simplify by the prime factors method. $\dfrac{x^8}{x^3}$

2. Simplify by the prime factors method. $-\dfrac{64n^6}{12n^4}$

3. Simplify using the quotient rule. $\dfrac{p^{13}}{p^8}$

4. Evaluate using the quotient rule. $\dfrac{12x^4}{4x^4}$

312

Name: _____   Date: _____

Instructor: _____   Section: _____

Notebook 23.2
Integer Exponents

Simplify using the prime factors method and the quotient rule.

$$\frac{x^3}{x^8} =$$

$$\frac{x^3}{x^8} =$$

What does a negative exponent mean?

Note: A negative exponent means _____.
It does not affect _____.

EXAMPLES 1–3 Simplify. Write your answer with positive exponents.

z^{-6}

$$\frac{x^3}{x^7}$$

$(x^{-5})(x^3)$

Remember the sign rules for raising a negative base to a power.

$(-a)^{even} =$ $(-a)^{odd} =$

$(-2)^2 =$ $(-2)^3 =$

EXAMPLES 4–6 Simplify. Write your answers with positive exponents.
2^{-5}

-5^{-2}

$(-3)^{-3}$

Assuming that m and n are real numbers, $x \neq 0$, and $y \neq 0$, what are the two properties of negative exponents?

$$\frac{1}{x^{-n}} =$$ $$\frac{x^{-m}}{y^{-n}}$$

EXAMPLES 7–9 Simplify. Write your answers with positive exponents.

$$\frac{1}{x^{-8}}$$

$$\frac{x^{-4}}{y^{-2}}$$

$$\frac{16^{-8}}{23^{-7}}$$

313

How do you know if an exponential expression is simplified?

all _____ have been removed,

all _____ are positive, and

all _____ are simplified.

What are the three properties for the Power Rules?

EXAMPLE 10 Simplify. Write your answer with positive exponents.

$(3x^{-4}y^2)^{-3}$

What does the Product Rule state?

EXAMPLE 11 Simplify. Write your answer with positive exponents.

$$\frac{x^2y^{-4}}{x^{-5}y^3}$$

What does Quotient Rule say?

EXAMPLE 11 Simplify. Write your answer with positive exponents.

$$\frac{(4x)^{-2}5y^{-7}}{x^6y^{-4}}$$

CONCEPT CHECK: Write $(ab)^2$ in simplest form.

GUIDED EXAMPLES:

Simplify. Write your answer with positive exponents.

1. m^{-16}

2. $\dfrac{a^{-2}}{a^5}$

Evaluate. Write your answers with positive exponents.

3. $(4a^3b^{-7})^{-4}$

4. $\dfrac{-7a^{-7}(6b^4)^{-2}}{a^{-2}b^{-11}}$

Name: _____   Date: _____

Instructor: _____   Section: _____

Notebook 23.3
Scientific Notation

When evaluating the powers of 10, a pattern emerges.
$10^0 =$
$10^1 =$
$10^2 =$
$10^3 =$
What does the positive exponent tell us?

$10^{-1} =$
$10^{-2} =$
$10^{-3} =$
What does the negative exponent tell us?

What is scientific notation?
Examples.

Write the steps for writing a number in scientific notation.
1.

2.

3.

EXAMPLES 1 & 2 Write in scientific notation.
23,400,000

0.0000034

How do you determine if the exponent is positive or negative?

Write the steps for writing a number in decimal notation.
1.

2.

3.

EXAMPLES 3 & 4 Write in decimal notation.
2.31×10^6

7.432×10^{-8}

Why is scientific notation useful?

EXAMPLE 5 A light-year is the distance that light travels in one year. This is a convenient unit to use when measuring distances in astronomy. One light-year is a distance of about 9,460,000,000,000,000 meters. Write this number in scientific notation.

EXAMPLE 6 Write the number in the following application in decimal notation.
The volume of a gold atom is 1.695×10^{-23} cubic centimeters. Write this in decimal notation.

Write the product rule:

Write the quotient rule:

EXAMPLES 7 & 8 Simplify.
$(8 \times 10^{16})(7 \times 10^{4})$

$\dfrac{1.5 \times 10^{7}}{2.5 \times 10^{2}}$

CONCEPT CHECK: If 9.123×10^{6} is changed to decimal notation, how many zeros will be in the final answer?

GUIDED EXAMPLES:

1. Write in scientific notation. 300,000,000

2. Write in scientific notation. 0.0000027

3. Write in decimal notation. 9.94×10^{5}

4. The diameter of a red blood cell is approximately 2.76×10^{-4} inches. Write this in decimal notation.

Notebook 23.4
Dividing a Polynomial by a Monomial

EXAMPLES 1 & 2 Simplify.

$$\frac{3}{5} + \frac{1}{5}$$

$$\frac{6x + 4}{2}$$

What is the rule for dividing a polynomial by a monomial?

EXAMPLE 3 Divide.

$$\frac{5x^5 + 10x^4 + 25x^3}{5x^2}$$

Split the division problem up.
Divide each term.
Write the sum of the results.

EXAMPLE 4 Divide.

$$\frac{63x^7 - 35x^6 - 49x^5}{-7x^3}$$

EXAMPLE 5 Divide.

$$(36x^3 - 18x^2 + 9x) \div (9x)$$

EXAMPLE 6 Divide.

$$\frac{16x^4 - 9x^3 + 24x^2}{12x^2}$$

317

EXAMPLE 7 Divide.

$$\frac{24x^3 + 16x^2 - 56x}{8x^2}$$

CONCEPT CHECK:

The final division for $\dfrac{50x^8 - 30x^5 + 20x^3}{10x^4}$ is $\dfrac{20x^3}{10x^4}$. How is this simplified?

GUIDED EXAMPLES:

1. Divide. $\dfrac{72x^5 + 36x^4 + 24x^3}{12x^2}$

2. Divide. $(170x^4 - 240x^3 + 70x^2) \div (10x)$

3. Divide. $\dfrac{-42x^9 + 25x^7 + 14x^5}{-7x^4}$

4. Divide. $\dfrac{80x^8 - 32x^5 + 64x^3}{16x^4}$

Notebook 23.5
Dividing a Polynomial by a Binomial

Review the procedure for long division.
1. Set up the problem using _____.
2.
3.
4.
5.
6.

Examples of Long Division

Write the steps for polynomial long division.
1. Set up the problem using _____.
2. Place the terms in _____.
   Insert a _____ for any missing terms.
3. Divide the _____ of the _____
   by the _____ of the _____.
4. _____ the first term of the answer by the _____.
5. _____ this product from the first two terms.
6. _____ the next term form the dividend.
7. Repeat this process until the _____ of the _____ is
   less than the _____.
8. If the divisor has a remainder, write the _____
   as a fraction over the _____, and add to the _____.

Note: In step 2, what do we mean by missing terms?

EXAMPLE 1 Divide. $(6x^2 + 7x + 2) \div (2x + 1)$

EXAMPLE 2 Divide. $(x^3 + 5x^2 + 11x + 4) \div (x + 2)$

EXAMPLE 3 Divide. $(5x^3 - 24x^2 + 9) \div (5x + 1)$
Don't forget your missing term!

CONCEPT CHECK:

What is the procedure if a polynomial long division problem has a remainder?

GUIDED EXAMPLES:

1. Divide. $(21x^2 + 34x + 8) \div (3x + 4)$

2. Divide. $(8x^3 - 30x^2 + 53x - 37) \div (4x - 5)$

3. Divide. $(y^3 - y - 24) \div (y - 3)$

4. Divide. $(n^3 - 1) \div (n - 1)$

Name: _____   Date: _____

Instructor: _____   Section: _____

Notebook 24.1
Greatest Common Factor

What are some factors of 6?
What is the factored form of 6?

When two or more numbers, variables, or algebraic expressions are multiplied, each is called a _____.

What does it mean "to factor"?

Note: What is factoring the reverse of?
How can you check your factoring?

What does it mean "to factor" a polynomial?

Factoring changes a sum/difference to a _____.

The _____ of two or more number is the largest number that divides exactly into each of the numbers.

Find the GCF for x^2 and x^3.

Write the procedure for finding the GCF of variable terms.

EXAMPLES 1–3 Find the GCF.
x^3, x^7, and x^5

y, y^4, and y^7

x and y^2

Now, write the procedure for finding the GCF of two or more terms with coefficients and variables.

EXAMPLE 4 Find the GCF. $9x^2$ and $15x^3$
GCF of numerical parts:
GCF of variable factors:
Product of these GCFs:

What does the Distributive Property state?

Write the steps for factoring a polynomial with common factors.

1.

2.

3.

4.

What is a prime polynomial?

EXAMPLE 5 Factor out the GCF. $9x^5 + 18x^2 + 3x$

GCF =

EXAMPLE 6 Factor out the GCF. $8x^3y + 16x^2y^2 - 24x^3y^3$

GCF =

EXAMPLES 7 & 8 Factor out the GCF.

$24ab + 12a^2 + 36a^3$

$3x + 7y + 12xy$

CONCEPT CHECK:

Why is x^3 the greatest common factor of the variable factors in $16x^5 - 2x^4 + 4x^3$?

GUIDED EXAMPLES:.

1. Find the GCF. $4x^3$ and $10x^5$

2. Find the GCF. $4ab^3$, $8a^4b^2$, and $6a^3b^3$

3. Factor by factoring out the GCF. $4x^4 + 12x^3 + 2x$

4. Factor by factoring out the GCF. $30x^3y^2 - 24x^2y^2 + 6xy^2$

Name: _____   Date: _____

Instructor: _____   Section: _____

Notebook 24.2
Factoring by Grouping

"To factor" means to write as a _____.
A common factor of a polynomial can be a _____,
a _____, or an algebraic _____.

Write the steps for factoring by grouping.
1.

2.

3.

4.

EXAMPLE 1 Factor by grouping. $2x^2 + 3x + 6x + 9$
Group terms.
Factor within groups.
Factor entire polynomial.
Multiply to check.

Note: Sometimes you will need to factor out a _____ to
obtain two terms that contain the _____ binomial factor.

EXAMPLE 2 Factor by grouping. $2x^2 + 5x - 4x - 10$
Group terms.
Factor within groups.
Factor entire polynomial.
Multiply to check.

Caution! When factoring out a _____, check carefully!!

EXAMPLE 3 Factor by grouping. $2ax - a - 2bx + b$

323

EXAMPLE 4 Factor by grouping. $10x^2 - 8xy + 15x - 12y$

CONCEPT CHECK:

What is the greatest common factor for each group of terms in the polynomial $(3x^2 + 6x) + (4x + 8)$?

GUIDED EXAMPLES:.

Factor by grouping.

1. $6x^2 - 15x + 4x - 10$

2. $4x + 8y + ax + 2ay$

3. $6xy + 14x - 15y - 35$

4. $3x + 6y - 5ax - 10ay$

324

Name: _____   Date: _____

Instructor: _____   Section: _____

Notebook 24.3
Factoring Trinomials of the form $x^2 + bx + c$

"To factor" means to write as a _____.

To multiply binomials, we can use the _____ method.
Example. Multiply $(x + 3)(x + 2)$.

Notes:
Where did the product of the first terms of the binomials end up?
Where did the product of the last two terms of the binomials go?
What about the sum of the last two terms of the binomials?

How do you factor trinomials of the form $x^2 + bx + c$?
1.

2.

3.

4.

EXAMPLE 1 Factor. $x^2 + 7x + 12$
Write the first two terms of factors.
List possible pairs of factors.
Find numbers whose product is last term and sum is middle term.
Write these numbers as the constants.
Check.

Caution! Be sure to include the correct signs.

EXAMPLE 2 Factor. $x^2 - 8x + 15$

Note: How will trinomials of the form $x^2 - bx + c$ factor?

EXAMPLE 3 Factor. $x^2 - 3x - 10$

325

EXAMPLE 4 Factor. $2x^2 + 24x + 40$

GCF =

Factor out the GCF first.

EXAMPLE 5 Factor. $4x^2 + 40x - 96$.

GCF =

Factor out the GCF first.

A _____ is a polynomial that cannot be factored.

EXAMPLE 6 Factor. $x^2 + 3x + 10$

CONCEPT CHECK:

Consider the polynomial $x^2 - 11x + 24$. Which factors of 24 will be the constants of the binomial factors?

GUIDED EXAMPLES:

Factor.

1. $x^2 + 8x + 12$

2. $x^2 - 11x + 18$

3. $x^2 - 5x - 24$

4. $2x^2 + 34x + 60$

Notebook 24.4
Factoring Trinomials of the Form $ax^2 + bx + c$

To multiply binomials, we can use the _____ method.
Example. Multiply using the FOIL method. $(2x + 3)(x + 1)$.

Notes:
What makes up the first term of the trinomial?
What makes up the last term of the trinomial?
What makes up the middle term of the trinomial?
This method of factoring is called _____.

How do you factor trinomials of the form $ax^2 + bx + c$?
1. List the different factorizations of _____.

2. List the different factorizations of _____.

3. List the possible combinations until the correct one is reached:
      a. What if c is positive?

      b. What if c is negative?

4. Check the _____ term.
      How do you check it?

EXAMPLE 1 Factor. $2x^2 + 5x + 3$
List factorizations of $2x^2$.
List factorizations of 3.
List possible combinations checking middle term.
Check.

EXAMPLE 2 Factor. $4x^2 - 21x + 5$
List factorizations of $4x^2$.
List factorizations of 5.
List possible combinations checking middle term.
Check.

EXAMPLE 3 Factor. $10x^2 - 9x - 9$

EXAMPLE 4 Factor. $3x^2 + 5x - 12$

CONCEPT CHECK:
Can $ax^2 + bx + c$ be factored the same way as $x^2 + bx + c$?

GUIDED EXAMPLES:
Factor.
1. $2x^2 + 7x + 5$

2. $2x^2 - 11x - 6$

3. $6x^2 + 5x - 25$

4. $10x^2 - 37x + 21$

328

Notebook 24.5
More Factoring of Trinomials by the Reverse FOIL Method

Recall the steps for factoring trinomials of the form $x^2 + bx + c$ and $ax^2 + bx + c$.

The _____ of a list of terms is the product of the GCF of the _____ and the CGF of the _____ factors.

First step: What do you do if a trinomial has a GCF other than 1?

EXAMPLE 1 Factor. $9x^2 + 3x - 30$
GCF??
Factor resulting trinomial.

Don't forget to write the GCF as part of your answer.

EXAMPLE 2 Factor. $3 - 10x + 8x^2$
Rearrange.
GCF??
Factor resulting trinomial.

What do you call a polynomial that cannot be factored?

EXAMPLE 3 Factor. $5x^2 + 7x + 4$
GCF??
Factor resulting trinomial.

EXAMPLE 4 Factor. $2x^2 + 12x + 24$

GCF??

Factor resulting trinomial.

Why is this one trinomial not prime?

CONCEPT CHECK:

What is the first step of factoring $2x^2 + 4x - 16$?

GUIDED EXAMPLES:

Factor.

1. $8x^2 + 8x - 6$

2. $24x^3 - 38x^2 + 10x$

3. $36x + 21x^2 + 3x^3$

4. $3x^3 - 15x^2 + 9x$

Name: _____   Date: _____

Instructor: _____   Section: _____

Notebook 24.6

Factoring Trinomials by Grouping Numbers (the *ac* Method)

OTHER NOTES

Write the steps for factoring a polynomial by grouping.

1. Factor out a _____.
2. Group _____.
3. Factor _____ groups.
4. _____ the entire polynomial.
5. Check.

When is factor by grouping primarily used?

Factor by grouping. $2x^2 + 3x + 6x + 9$

We can also factor trinomials by this grouping method.

Write the steps for factoring trinomials by grouping.

1. Factor out a _____, if there is one.
2. Multiply __ and __.
3. Find factors of ____ that add to _____.
4. Rewrite the trinomial, and replace ___ with the
   _____ found in step 3.
5. Factor by _____.

EXAMPLE 1 Factor by the *ac*-method. $x^2 + 7x + 12$

GCF??

Multiply to find *ac*.

Find factors of *ac* that add to *b*.

Rewrite as 4 terms.

Factor by grouping.

EXAMPLE 2 Factor by the *ac*-method. $8x^2 - 19x + 6$

GCF??

Multiply to find *ac*.

Find factors of *ac* that add to *b*.

Rewrite as 4 terms.

Factor by grouping.

EXAMPLE 3 Factor by the *ac*-method. $18x^3 - 33x^2 - 21x$
GCF??
Multiply to find *ac*.
Find factors of *ac* that add to *b*.
Rewrite as 4 terms.

What happens if none of the factors of *ac* add up to *b*?

EXAMPLE 4 Factor by the *ac*-method. $3x^2 - 6x - 5$

EXAMPLE 5 Factor by the *ac*-method. $6x^2 + 11xy + 4y^2$

CONCEPT CHECK:
How should the expression $8x^2 - 10x + 3$ be rewritten with 4 terms?

GUIDED EXAMPLES: See guided examples for the specific problems.
Factor.

1. Factor. $8x^2 + 8x - 6$

2. Factor. $2x^2 + 5xz - 12z^2$

3. Factor. $18x^6 - 84x^3 + 48$

4. Factor. $4x^2 + 10x + 3$

Notebook 25.1
Special Cases of Factoring

The product of a sum and a difference of binomials
$(a + b)(a - b) =$ _____, where a and b are
numbers or algebraic expressions.

EXAMPLE 1 Factor. $x^2 + 0x - 9$

How does a difference of two squares factor?

Example $x^2 - 9$

What are the 3 steps in identifying a difference of two squares?
1. The expression is a _____,
2. Both terms are _____,
3. The terms are _____.

EXAMPLES 2–4 Determine if each expression is a
difference of two squares. Tell why or why not?
$x^2 - 16$

$x^2 - 7$

$4x^2 + 81$

List the first 20 perfect square numbers:

Write the steps for factoring a difference of two squares.
1.

2.

3.

When will a sum of two squares factor?

EXAMPLES 5 & 6 Factor. Remember, $a^2 - b^2 = (a + b)(a - b)$.
$x^2 - 49$

$25b^2 - 64$

OTHER NOTES

EXAMPLES 7 & 8 Factor.

$4x^2 - 81y^2$

$-9x^2 + 1$

A _____ is equal to $(a + b)^2 = a^2 + 2ab + b^2$, or $(a - b)^2 = a^2 - 2ab + b^2$, where a and b are numbers or expressions.

What are the 2 steps in identifying a perfect square trinomial?
1. The first and last terms are _____,
2. The middle term is _____.

To factor a perfect square trinomial one of the formulas below.
$a^2 + 2ab + b^2 = (a + b)^2$ or $a^2 - 2ab + b^2 = (a - b)^2$.

EXAMPLES 9 & 10 Factor each of the perfect square trinomials completely.

$x^2 + 6x + 9$ $\qquad\qquad\qquad$ $9n^2 - 66n + 121$

Note: What does the middle term of the trinomial tell you about the signs in the binomial factors?

Caution! Always look for a _____ first.

EXAMPLES 11 & 12 Factor.

$25x^2 - 100$ $\qquad\qquad\qquad$ $100x^2 + 400x + 400$

CONCEPT CHECK:
Which characteristic disqualifies $x^2 + 25$ from being a difference of squares?

GUIDED EXAMPLES:
1. Determine if the polynomial is a difference of two squares. $49r^2 - 10$

2. Factor completely. $x^2 - 121$

3. Factor completely. $x^2 + 28x + 196$

4. Factor completely. $9x^2 - 72x + 144$

Name: _____   Date: _____

Instructor: _____   Section: _____

Factoring Polynomials

                                                                    OTHER NOTES

Factoring – A General Strategy
To factor a polynomial,
1. Look for a _____. If there is one, factor out the GCF.

2. Look at the _____ of terms.
      a. Two terms: Is it a _____?
         What is the factored form of $a^2 - b^2$?
      b. Three terms: Is it a _____?
         What is the factored form of $a^2 + 2ab + b^2$?
         What is the factored form of $a^2 - 2ab + b^2$?
      c. Otherwise, _____ by any method.
      d. Four terms: Try to _____.

3. Always _____ by making sure all
   common factors are factored and each factor is _____.

4. Check by _____.

EXAMPLE 1 Factor completely. $3k^2 - 48$

EXAMPLE 2 Factor completely. $12n^2 - 12n - 144$

EXAMPLE 3 Factor completely. $4x^2 - 20x + 25$

EXAMPLE 4 Factor completely. $x^3 + 2x^2 - 9x - 18$

EXAMPLE 5 Factor completely. $6b^4 - 15b^2 - 36$

EXAMPLE 6 Factor completely. $12n^2 - 17n - 7$

CONCEPT CHECK:
Can $(x^2 + 4)(x^2 - 4)$ be factored further?

GUIDED EXAMPLES:
Factor completely.
1. $6x^2 - 486$

2. $10x^4 - 75x^3 + 135x^2$

3. $-2x^3 + 10x^2 + 72x - 360$

4. $6h^8 - 6$

Notebook 25.3
Factoring the Sum and Difference of Cubes

A difference of two squares can be factored according to the pattern of
$a^2 - b^2 = $ _____, where a and b are numbers or algebraic
expressions.

Can a sum of two squares factor?

Examples of differences of two squares
$x^2 - 25$

$9x^2 - 64$

$36 - x^2$

Write some examples of perfect cubes.

You will need this list of perfect cubes to identify a sum or difference of two
cubes.

How do you identify a sum or difference of two cubes?

EXAMPLES 1–3 Determine if each expression is a sum of two
cubes, a difference of two cubes, or neither. Tell why or why not?
$x^3 - 27$

$x^3 + 75$

$125x^6 + y^{12}$

To factor a sum of two cubes, use the formula $a^3 + b^3 = \dots$

To factor a difference of two cube, use the formula $a^3 - b^3 = \dots$
Caution! Pay special attention to the _____ in the factors.

Write the steps for factoring a sum or difference of two cubes.
1. Factor out any _____.

2. Write each term as a _____.

3. Write the _____ using the properties above.
Note: Will the trinomial factor ever factor?

EXAMPLE 4 Factor completely. $x^3 + 125$.

EXAMPLE 5 Factor completely. $x^3 - 64$

Caution! What is the first step in factoring??

EXAMPLES 6 & 7 Factor completely.
$3x^2 + 3$

$216x^2 - 64y^3$

CONCEPT CHECK:
Write the formula that can be used to factor $x^3 - 512$.

GUIDED EXAMPLES:
1. Determine if the polynomial is a sum of two cubes, a difference of two cubes, or neither. $27r^3 + 512$

2. Factor completely. $y^3 - 343$

3. Factor completely. $27x^3 + 729$

4. Factor completely. $2x^3 - 32$

338

Notebook 25.4
Solving Quadratic Equations by Factoring

What is the standard form of a quadratic equation?
What is the highest degree of any term in a quadratic equation?

What does the Zero Property of Multiplication state?

What is the procedure for solving a quadratic equation by factoring?
1. Make sure the equation is in _____.

2. _____, if possible.

3. Set each factor _____.

4. _____ the resulting equation.

5. _____ each solution in the original equation.

What is the most number of solutions a quadratic equation can have?
What is a double root?
When might a double root occur?

EXAMPLE 1 Solve by factoring. $x^2 + 4x = 0$
Factor.
Set each factor = 0.
Solve.
Check.

EXAMPLE 2 Solve by factoring. $2x^2 - 8x - 42 = 0$

EXAMPLE 3 Solve by factoring. $10x^2 - x = 2$
Write in standard form (set = 0)
Factor.
Set each factor = 0.
Solve.
Check.

EXAMPLE 4 Solve by factoring. $-5t^2 + 13t + 6 = 0$
Write in standard form (set = 0).
Factor.
Set each factor = 0.
Solve.
Check.

EXAMPLE 5 Solve by factoring. $4x^2 + 9 = 12x$

EXAMPLE 6 Solve by factoring. $x^2 - 64 = 0$

CONCEPT CHECK:
An equation has been factored to $(x + 7)(x + 6) = 0$. What are the solutions?

GUIDED EXAMPLES:
Solve by factoring.
1. $6x^2 - 12x = 0$

2. $4x^2 - 4x - 120 = 0$

3. $16x^2 = 56x - 49$

4. $-x^2 - 6x + 27 = 0$

Notebook 25.5
Applications of Quadratic Equations

EXAMPLE 1 A cliff diver jumps from a platform placed on a cliff approximately 144 feet above the surface of the sea. Disregarding air resistance, the height S, in feet, of a cliff diver above the ocean after t seconds is given by the quadratic equation $S = -16t^2 + 144$. How long does it take the diver to reach the water? (Note: The height when he hits the water is 0 feet.) Set up the equation.

Solve by factoring.

Why ignore $t = -3$?

What is the quadratic equation that gives the approximate height of an object that is thrown or launched upward?

$S =$

$v =$

$t =$

$h =$

EXAMPLE 2 A tennis ball is thrown upward with an initial velocity of 8 meters per second. Suppose that the initial height above the ground is 4 meters. (What formula will you use?)

Find the height S of the ball after 1 second.

At what time t will the ball hit the ground?

Why eliminate $t = -\dfrac{2}{5}$?

EXAMPLE 3 A rocket is fired upward with an initial velocity (v) of 110 meters per second. The quadratic equation $S = -5t^2 + 110t$ can be used to find the height (S) of the rocket, in meters, at any time (t) in seconds.

Find the height of the rocket 10 seconds after it takes off ($t = 10$).

Find how many seconds it takes for the rocket to reach a height of 425 meters.

EXAMPLE 4 The length of the base of a rectangle is 7 inches greater than the height. If the total area of the rectangle is 120 square inches, what are the lengths of the base and height of the rectangle? Remember that the area for a rectangle can be found using the formula $A = BH$.

Draw a picture, label the base and height.

Write an equation.

Solve and state the answer.

CONCEPT CHECK: Priscilla is solving an equation for the length of time that it takes a diver to reach the water. She solves $S = -16t^2 + 128t + 144$ for t by factoring and finds two answers. She ignores one answer. Why might she do this?

GUIDED EXAMPLES:

1. A ball is thrown upward with an initial velocity v of 13 meters per second. Suppose that the initial height h above the ground is 6 meters. At what time t will the ball hit the ground? The ball is on the ground when $S = 0$. Use the equation $S = -5t^2 + vt + h$.

2. A rocket is fired upward with a velocity v of 480 feet per second. Find how many seconds it takes for the rocket to reach a height of 3200 feet. Use the quadratic equation $S = -16t^2 + vt$.

3. The length of the base of a rectangle is 3 inches less than twice the height. If the total area of the rectangle is 152 square inches, what are the lengths of the base and height of the rectangle? Remember that for a rectangle Area = base · height.

Notebook 26.1
Undefined Rational Expressions

OTHER NOTES

Any _____ can be written as a fraction of two integers.

What is a rational expression?

Give some examples of rational expressions.

Note: Do rules that work for numerical fractions work for rational expressions? Why or why not?

What does it mean if a rational expression is "undefined"?

EXAMPLE 1 Find any values of the variable that will make the rational expression undefined. $\dfrac{15x^2 + 25x}{5x}$

Set denominator = 0.

Solve.

EXAMPLE 2 Find the value of the variable that will make the rational expression undefined. $\dfrac{24x^2 + 9x}{8x + 3}$

Set denominator = 0.

Solve.

What does the Zero Property of Multiplication state?

EXAMPLE 3 Find any values of the variable that will make the rational expression undefined. $\dfrac{x^3 - 9x}{x^2 + 5x + 6}$

Set denominator = 0.

Solve.

EXAMPLE 4 Find any values of the variable that will make the rational expression undefined. $\dfrac{x^2+6x+8}{x^2-16}$

EXAMPLE 5 Find any values of the variable that will make the rational expression undefined. $\dfrac{x^3+2x^2+x+2}{x^2+1}$

EXAMPLE 6 The cost to make x dozen cookies can be modeled by the expression $\dfrac{3x+45}{4x}$. Evaluate this expression at $x=3$ and $x=0$. Discuss what each result means in the context of the application.
Understand the problem
Create a plan
Find the answer
Check the answer

EXAMPLE 7 The total revenue, R, in thousands of dollars, from the sale of a popular book, is given by the expression $\dfrac{1800x^2}{x^2+4}$, where x is the number of years since publication. Find the domain of this expression. Then find the total sales revenue of the book two years after it is published. .
Understand the problem
Create a plan
Find the answer
Check the answer

CONCEPT CHECK: Which values will make the function $f(x)=\dfrac{x+2}{x-1}$?

GUIDED EXAMPLES: Find any values of the variable that will make the rational expression undefined.

1. $\dfrac{51x^2+34x}{17x}$ 2. $\dfrac{72x^2+56x}{9x-7}$ 3. $\dfrac{8x+72}{x^2-81}$

4. The yearly electrical cost to run a certain television model with a purchase price of \$325 is given by the expression $\dfrac{325+13x}{x}$ where x is the number of years after purchase. What is the domain of this expression? What is the yearly cost to run the television four years after purchase?

Notebook 26.2

Simplifying Rational Expressions

Write the basic rule of fractions in terms of how they can be simplified.

Examples.

$$\frac{14}{21} \qquad \frac{15x}{13x} \qquad \frac{x^2+2x-3}{x^2-5x+4}$$

Caution! Only _____ can divide, not terms.

Example.

$$\frac{7+2}{1+2}$$

Which method will we use to simplify fractions?

EXAMPLE 1 Simplify. $\dfrac{21}{39}$

Factor the num. and den.

Divide by common factors.

Write the steps for simplifying rational expressions.

1.

2.

3.

EXAMPLE 2 Simplify. $\dfrac{2x+6}{3x+9}$

Factor the num. and den.

Divide by common factors.

EXAMPLE 3 Simplify. $\dfrac{x^2+9x+14}{x^2-4}$

Factor the num. and den.

Divide by common factors.

For all polynomials A and B, where $A \neq B$, $\dfrac{A-B}{B-A} = -1.$

345

EXAMPLE 4 Simplify. $\dfrac{2-x}{x-2}$

Factor the num. and den.
Divide by common factors.

EXAMPLE 5 Simplify. $\dfrac{5x^2-45}{45-15x}$

Factor the num. and den.
Divide by common factors.

EXAMPLE 6 Simplify. $\dfrac{x^2-7xy-12y^2}{2x^2-9xy+4y^2}$

Factor the num. and den.
Divide by common factors.

CONCEPT CHECK:
What is the basic rule of fractions?

GUIDED EXAMPLES:

1. Simplify by using the prime factors method. $\dfrac{105}{42}$

2. Simplify. $\dfrac{7x+42}{5x+30}$

3. Simplify. $\dfrac{3x-7}{14-6x}$

4. Simplify. $\dfrac{20-x-x^2}{3x^2-48}$

Notebook 26.3
Multiplying Rational Expressions

To multiply two fractions,
1. Find the _____ in the numerators and the denominators.
    Do not _____ yet.
2. _____ the numerators and denominators by
    _____.

3. Write the remaining factors as one _____.

EXAMPLE 1 Multiply. $\dfrac{12}{7} \cdot \dfrac{49}{36}$

Write the steps for multiplying rational expressions.
1. _____ the numerator and _____.
2. Find the _____. Do not multiply.
3. _____ the numerators and denominators
    by _____.
4. Write the _____ as one fraction.

What does the Quotient Rule say?

EXAMPLE 2 Multiply. $\dfrac{2x^2}{3y} \cdot \dfrac{6y}{8x}$

EXAMPLE 3 Multiply. $\dfrac{7x}{22} \cdot \dfrac{11x+33}{7x+21}$

Factor the num. and den.
Find common factors.
Divide by common factors.
Write as one fraction.

347

EXAMPLE 4 Multiply. $\dfrac{2x-14}{5x+25} \cdot \dfrac{x^2+7x+10}{4x-12}$

Factor the num. and den.
Find common factors.
Divide by common factors.
Write as one fraction.

Remember, $\dfrac{a-b}{b-a} = $ _____.

EXAMPLE 5 Multiply. $\dfrac{x^2-x-12}{16-x^2} \cdot \dfrac{2x^2+7x-4}{x^2-4x-21}$

EXAMPLE 6 Multiply. $\dfrac{3x^2+7x+2}{14x+21} \cdot \dfrac{7x-14}{5x^2+6x-8}$

CONCEPT CHECK:

What are the common factors when multiplying $\dfrac{2x^2}{y} \cdot \dfrac{6}{x}$?

GUIDED EXAMPLES:

1. Multiply. $\dfrac{-45x^3}{26y} \cdot \dfrac{39y}{30x^2}$

2. Multiply. $\dfrac{8x}{9} \cdot \dfrac{6x-36}{8x-48}$

3. Multiply. $\dfrac{x^2-5x-6}{x^2-2x-3} \cdot \dfrac{4x^2+17x-42}{x^2-36}$

4. Multiply. $\dfrac{x^4-1}{3x^2+3} \cdot \dfrac{12x^2+6x}{5x^3+8x^2+3x}$

Name: _____   Date: _____

Instructor: _____   Section: _____

Notebook 26.4
Dividing Rational Expressions

To divide two fractions,
1. Find the _____ of the second fraction
   and _____ the first fraction by this _____.
2. Find the _____ in the numerators
   and denominators. Do not multiply yet. _____.
3. _____ by common _____.
4. Write the remaining factors as one _____.

EXAMPLE 1 Divide. $\dfrac{12}{7} \div \dfrac{36}{49}$

Write the steps for dividing rational expressions.
1. Find the _____ of the second expression and
   _____ the first expression by this _____.

2. Factor the _____. Do not multiply.
3. _____ the numerators and denominators
   by _____.
4. Write the _____ as one fraction.

EXAMPLE 2 Divide. $\dfrac{-12x^2}{5y} \div \dfrac{18x}{15y}$

Multiply by the reciprocal.

EXAMPLE 3 Divide. $\dfrac{8x}{14} \div \dfrac{8x-32}{7x-28}$

Multiply by the reciprocal.

349

EXAMPLE 4 Divide. $\dfrac{x^2 + 3x - 10}{x^2 + x - 20} \div \dfrac{x^2 + 4x + 3}{x^2 - 3x - 4}$

Remember, $\dfrac{a-b}{b-a} =$ _____

EXAMPLE 5 Divide. $\dfrac{x-5}{3} \div (25 - x^2)$

CONCEPT CHECK:

Rewrite $\dfrac{x-2}{4} \div (x^2 - 4)$ as a multiplication problem.

GUIDED EXAMPLES:

1. Divide. $\dfrac{9x^3}{35y^2} \div \dfrac{27x}{7y}$

2. Divide. $\dfrac{16x^2 - 6x - 7}{24x^2 - 29x + 7} \div \dfrac{10x^2 + x - 2}{15x^2 + x - 2}$

3. Divide. $\dfrac{6x-7}{4} \div (49 - 36x^2)$

4. Divide. $\dfrac{3x+2}{x+4} \cdot \dfrac{x^2 - 16}{4x+3}$

Name: _____   Date: _____

Instructor: _____   Section: _____

Notebook 27.1
Adding Like Rational Expressions

A _____ can be written as a fraction of two algebraic expressions.

What are like rational expressions?

Write some examples of like rational expressions.

What is the rule for adding like rational expressions?

Write the steps for adding like rational expressions.
1.

2.

EXAMPLE 1 Add. $\dfrac{5a}{4a+2b} + \dfrac{6a}{4a+2b}$

Add the numerators.
Keep the denominator.
Simplify.

Why do you not divide the a from the answer?

EXAMPLE 2 Add. $\dfrac{-7m}{2n} + \dfrac{m}{2n}$

Add the numerators.
Keep the denominator.
Simplify.

Note: Simplify whenever possible by _____,
_____, and _____ common factors.

EXAMPLE 3 Add. $\dfrac{-3}{x^2-3x+2} + \dfrac{x+1}{x^2-3x+2}$

EXAMPLES 4 & 5 Add.

$$\frac{x+3}{x^2-1}+\frac{x}{x^2-1}$$

$$\frac{x}{x+1}+\frac{1}{x+1}$$

CONCEPT CHECK:

Is $\dfrac{2x+2}{x+1}$ written in simplest form?

GUIDED EXAMPLES:

1. Add. $\dfrac{-4b}{5a+b}+\dfrac{b}{5a+b}$

2. Add. $\dfrac{p}{12q}+\dfrac{2p}{12q}$

3. Add. $\dfrac{2x^2-8x}{2x-5}+\dfrac{x+5}{2x-5}$

4. Add. $\dfrac{2x}{x^2+5x+6}+\dfrac{x+2}{x^2+5x+6}$

352

Name: _____   Date: _____

Instructor: _____   Section: _____

Notebook 27.2
Subtracting Like Rational Expressions

OTHER NOTES

A rational expression can be written as a fraction of two

_____.

Rational expressions with a common denominator are called

_____.

To add like rational expressions.
1. _____ the numerators and _____ the denominator.
2. _____ if possible.

What does "Leave, Change, Change" mean?

What is the rule for subtracting like rational expressions?

Write the steps for subtracting like rational expressions.
1.

2.

EXAMPLE 1 Subtract. $\dfrac{-2a}{3b} - \dfrac{5a}{3b}$

Subtract the numerators.
Keep the denominator.
Simplify.

EXAMPLE 2 Subtract. $\dfrac{8x}{2x+3y} - \dfrac{3x}{2x+3y}$

Subtract the numerators.
Keep the denominator.
Simplify.

Why is the x not divided out of the answer?

Note: Be sure to treat the numerator of the fraction
being subtracted as a single quantity. How do you do this?

EXAMPLE 3 Subtract. $\dfrac{3x^2+2x}{x^2-1}-\dfrac{10x-5}{x^2-1}$

EXAMPLES 4 & 5 Subtract.

$$\dfrac{3x}{x-4}-\dfrac{12}{x-4}$$

$$\dfrac{3x}{x^2+3x+2}-\dfrac{2x-8}{x^2+3x+2}$$

CONCEPT CHECK:

Is $\dfrac{2c-4}{8c}$ written in simplest form?

GUIDED EXAMPLES:

1. Subtract. $\dfrac{5x}{2x-10}-\dfrac{6}{2x-10}$

2. Subtract. $\dfrac{10m}{2m+n}-\dfrac{7m+4}{2m+n}$

3. Subtract. $\dfrac{-8a}{3b}-\dfrac{a}{3b}$

4. Subtract. $\dfrac{2y}{2y-7}-\dfrac{7}{2y-7}$

Name: _____   Date: _____

Instructor: _____   Section: _____

Finding the Least Common Denominator for Rational Expressions

OTHER NOTES

To add and subtract rational expressions when the denominators are not the same, first find the _____.

What is the LCD?

Write the steps for finding the LCD of two or more expressions.
1. _____ each denominator.
2. List each _____.
3. For each factor, put the _____ that appears on each factor.
4. The LCD is the _____ of these _____.

Note: If a factor occurs more than once in any one denominator, how many times does the LCD contain that factor?

EXAMPLE 1 Find the LCD. $\dfrac{5}{2x-4}, \dfrac{6}{3x-6}$

Factor each denominator.
List each different factor.
Put the highest exponent on each factor.
LCD =

EXAMPLE 2 Find the LCD. $\dfrac{5}{12ab^2c}, \dfrac{13}{18a^3bc^4}$

Factor each denominator.
List each different factor.
Line up the factors.
Put the highest exponent on each factor.
LCD =

Caution! Be careful when comparing common _____.

EXAMPLES 3–5 Find the LCD.

$$\frac{5}{x+3}, \frac{2}{x-4}$$

$$\frac{8}{x^2-5x+4}, \frac{12}{x^2+2x-3}$$

$$\frac{x+3}{x^2-6x+9}, \frac{10}{2x^2-4x-6}, \frac{x}{2}$$

CONCEPT CHECK:

How many times will $(x-1)$ be included in the LCD

for $\dfrac{x}{(x-1)^2}, \dfrac{1}{x-1}$?

GUIDED EXAMPLES:

1. Find the LCD. $\dfrac{7}{6x+21}, \dfrac{13}{10x+35}$

2. Find the LCD. $\dfrac{3}{50xy^2z}, \dfrac{19}{40x^3yz}$

3. Find the LCD. $\dfrac{2}{x^2+5x+6}, \dfrac{6}{3x^2+5x-2}$

4. Find the LCD. $\dfrac{x-3}{x^3+8x^2+16x}, \dfrac{7x}{6x^2+18x-24}, \dfrac{5}{x}$

Name: _____   Date: _____

Instructor: _____   Section: _____

Notebook 27.4
Adding and Subtracting Unlike Rational Expressions

To add or subtract _____ fractions, first rewrite each fraction using the
_____. Then _____ or _____, and simplify, if possible.

Write the steps for writing equivalent rational expressions with a least common
denominator.
1. Find the _____.

2. For each rational expression, find the _____ that you
   need to _____ the denominator by to get the _____.

3. Rewrite each rational expression as an _____
   rational expression whose _____ is the LCD.
   For each rational expression, multiply the _____
   and _____ by the expression found in step 2.

EXAMPLE 1 Write equivalent rational expressions using the LCD.

$\dfrac{8}{ab}, \dfrac{5}{a^2}$ LCD =

Write the steps for adding and subtracting rational expressions.
1. If the rational expressions have _____ denominators,
   find the _____. Otherwise, go straight to step 3.

2. _____ each rational expression with the LCD
   as the _____.

3. Add or subtract the _____, and keep the _____.

4. _____ if possible.

EXAMPLE 2 Add. $\dfrac{5}{xy} + \dfrac{2}{y}$

LCD =
Rewrite as equivalent fractions.
Add numerators.
Keep denominator.

357

EXAMPLE 3 Add. $\dfrac{3x}{x^2 - y^2} + \dfrac{5}{x+y}$

LCD =
Equivalent fractions.
Add numerators.
Keep denominator.
Simplify.

EXAMPLE 4 Add. $\dfrac{3y}{y^2 + 4y + 3} + \dfrac{2}{y+1}$

Caution! Why is it helpful to place parentheses around the numerator of the second fraction when subtracting?

EXAMPLE 5 Subtract. $\dfrac{3x-4}{x-2} - \dfrac{5x-6}{2x-4}$

EXAMPLE 6 Subtract. $\dfrac{-3}{x^2 + 8x + 15} - \dfrac{1}{2x^2 + 7x + 3}$

CONCEPT CHECK:

What is the LCD for the following sum, $\dfrac{3}{x} + \dfrac{2}{x-1}$?

GUIDED EXAMPLES:

1. Add. $\dfrac{7}{a} + \dfrac{3}{ab}$

2. Add. $\dfrac{7x}{x^2 + 2xy + y^2} + \dfrac{4}{x^2 + xy}$

3. Subtract. $\dfrac{x+7}{3x-9} - \dfrac{x-6}{x-3}$

4. Subtract. $\dfrac{x}{x^2 + 2x - 3} - \dfrac{x}{x^2 - 5x + 4}$

Name: _____   Date: _____

Instructor: _____   Section: _____

Notebook 28.1

Simplifying Complex Rational Expressions by Adding and Subtracting

What is a complex rational expression?

Write some examples of complex rational expressions.

Note: What does the fraction bar in a complex fraction mean?

To add or subtract fractions, you must have a _____.

For which type of complex fraction problem does this method of simplifying a complex rational expression work best for?

Write the steps for simplifying a complex rational expressions by inverting and multiplying.

1. If necessary, simplify the _____ into a single fraction.
2. If necessary, simplify the _____ into a single fraction.
3. _____ the fraction in the _____ by the _____ in the _____. To do this, _____ the fraction in the _____ and _____ it by the _____.
4. _____, if possible.

EXAMPLE 1 Simplify. $\dfrac{\frac{1}{x}}{\frac{2}{y^2}+\frac{1}{y}}$

True/False A complex rational expression may contain two or more fractions in the numerator and the denominator.

EXAMPLE 2 Simplify. $\dfrac{\frac{1}{x}+\frac{1}{y}}{\frac{3}{x}-\frac{2}{y}}$

Numerator as a single fraction.
Denominator as a single fraction.
Invert and multiply.

359

EXAMPLE 3 Simplify. $\dfrac{\frac{1}{x^2-1}+\frac{2}{x+1}}{x}$

Numerator as a single fraction.
Denominator as a single fraction.
Invert and multiply

EXAMPLE 4 Simplify. $\dfrac{\frac{3}{a+b}-\frac{3}{a-b}}{\frac{5}{a^2-b^2}}$

CONCEPT CHECK:
Write an example of a complex fraction with two fractions
in the numerator and two in the denominator?

GUIDED EXAMPLES:

1. Simplify. $\dfrac{\frac{1}{a}+\frac{1}{a^2}}{\frac{2}{b^2}}$

2. Simplify. $\dfrac{\frac{1}{x}+\frac{1}{y}}{\frac{x}{2}-\frac{5}{y}}$

3. Simplify. $\dfrac{\frac{x}{x^2+4x+3}+\frac{2}{x+1}}{x+1}$

4. Simplify. $\dfrac{\frac{6}{x^2-y^2}}{\frac{1}{x-y}+\frac{3}{x+y}}$

Name: _____   Date: _____

Instructor: _____   Section: _____

Notebook 28.2
Simplifying Complex Rational Expressions by Multiplying by the LCD

OTHER NOTES

A _____ is a rational expression that contains a fraction in the numerator, in the denominator, or in both.

Write some examples of complex rational expressions.

Note. What two things can the fraction bar in a complex fraction act as?

If a is a real number and $a \neq 0$, then what effect does multiplying a fraction by $\dfrac{a}{a}$ have?

Write the steps for simplifying a complex rational expressions by multiplying by the least common denominator.
1. Determine the _____ of all denominators in the _____.
2. _____ both the numerator and the denominator by the ____.
3. _____, if possible.

For which type of complex fraction problem does this method of simplifying a complex rational expression work best for?

EXAMPLE 1 Simplify by multiplying by the LCD. $\dfrac{\frac{3}{x}}{\frac{2}{x^2}+\frac{5}{x}}$

Determine the LCD of all denominators.
Multiply num and den by the LCD.
Simplify.

EXAMPLE 2 Simplify by multiplying by the LCD. $\dfrac{\frac{5}{ab^2}-\frac{2}{ab}}{3-\frac{5}{2a^2b}}$

LCD =

EXAMPLE 3 Simplify by multiplying by the LCD. $\dfrac{\frac{3}{a+b} - \frac{3}{a-b}}{\frac{5}{a^2-b^2}}$

LCD =

CONCEPT CHECK:

Write an example of a complex fraction with two fractions in the numerator and two in the denominator?

GUIDED EXAMPLES:

1. Simplify by multiplying by the LCD. $\dfrac{\frac{1}{a} + \frac{2}{b}}{\frac{3}{ab}}$

2. Simplify by multiplying by the LCD. $\dfrac{\frac{2}{3x^2} - \frac{3}{y}}{\frac{5}{xy} - 4}$

3. Simplify by multiplying by the LCD. $\dfrac{\frac{6}{x^2-y^2}}{\frac{1}{x-y} + \frac{3}{x+y}}$

4. Simplify by multiplying by the LCD. $\dfrac{\frac{x}{x+2} + \frac{4}{x^2+6x+8}}{\frac{1}{x+1}}$

362

Notebook 28.3
Solving Rational Equations

The _____ states that if both sides of an equation are multiplied by the same non-zero number, the solution does not change.

Write the steps for solving an equation containing rational expressions.
1. Find the _____ of all the denominators.
2. _____ each term of the equation by the _____.
3. _____, then solve the resulting equation.
4. Check. Any value that would make a _____ equal to _____ must be _____ from the solution.

EXAMPLE 1 Solve for x. $\dfrac{5}{x} + \dfrac{2}{3} = -\dfrac{3}{x}$.

LCD =
Multiply each term by the LCD.
Solve.

EXAMPLE 2 Solve for x. $\dfrac{6}{x+3} = \dfrac{3}{x}$

LCD =

EXAMPLE 3 Solve for x. $\dfrac{3}{x+5} - 1 = \dfrac{4-x}{2x+10}$.

LCD =

What is an extraneous solution?

Why is it important that you check all apparent solutions in the original equation?

How many solutions will there be?

EXAMPLE 4 Solve for x. $\dfrac{y}{y-2} - 4 = \dfrac{2}{y-2}$

LCD =

Make sure you check your solution. What happens?

CONCEPT CHECK:

What is the first step in solving a rational equation?

GUIDED EXAMPLES:

1. Solve for x. $\dfrac{6}{3x-5} = \dfrac{3}{2x}$

2. Solve for x. $\dfrac{3}{x+3} + \dfrac{9}{x^2+3x} = \dfrac{x-2}{x}$

3. Solve for x. $\dfrac{2x}{x+1} = \dfrac{-2}{x+1} + 1$

4. Solve for x. $\dfrac{x+11}{x^2-5x+4} + \dfrac{3}{x-1} = \dfrac{5}{x-4}$

Name: _____   Date: _____

Instructor: _____   Section: _____

Notebook 28.4
Applications of Rational Equations: Solving Formulas for a Variable

To solve an equation containing rational expressions:
1. Find the _____ of all the denominators.
2. _____ each term of the equation by the _____.
3. _____, then _____ the resulting equation.
4. Check. Any value that would make a _____ equal to _____ must
   be _____ from the solution.

What is the difference between solving an equation and solving a formula?

EXAMPLE 1 The following formula is used in the study of light
passing through a lens. It relates the focal length f of the lens to
the distance a of an object from the lens and the distance b of the
image from the lens. Solve for the variable a. $\dfrac{1}{f} = \dfrac{1}{a} + \dfrac{1}{b}$

LCD =
Multiply each term by the LCD.
Solve.

EXAMPLE 2 The gravitational force F between two masses m_1 and
m_2 a distance d apart is represented by the following formula. In
this equation, G is the gravitational constant. Solve for m_2. $F = \dfrac{Gm_1m_2}{d^2}$

365

EXAMPLE 3 The slope of a line that passes through the points (x_1, y_1) and (x_2, y_2) is shown below. Solve for x_1. $m = \dfrac{y_2 - y_1}{x_2 - x_1}$

LCD =

CONCEPT CHECK:

For the formula $a = \dfrac{x_1 + x_2 + x_3}{n}$, what operations are necessary to solve the formula for the variable x_3?

GUIDED EXAMPLES:

1. The following formula relates the total amount of time t in hours that is required for two workers to complete a job working together if one worker can complete it alone in c hours and the other worker in d hours.

   Solve for t. $\quad \dfrac{1}{t} = \dfrac{1}{x} + \dfrac{1}{d}$

2. The number of telephone calls C between two cities of populations p_1 and p_2 that are a distance d apart may be represented by the formula below, where B is a constant. Solve this equation for p_1. $\quad C = \dfrac{Bp_1 p_2}{d^2}$

3. The formula $Q = \dfrac{kA(t_1 - t_2)}{L}$ is used to measure heat conduction. In this formula, Q is the rate of conduction, k is the thermal conductivity of the material, A is the cross-sectional area, L is the length, and $(t_1 - t_2)$ is the difference in time. Solve for t_1.

Notebook 28.5
Applications of Rational Equations: Work Problems

If a task can be completed in t hours, then _____ of the task can be completed in one hour.

EXAMPLES 1 – 3 If Alisha can complete the following tasks in the following times, how much of the task can she complete in an hour? In other words, what is her rate at completing the task?

Mow the lawn: 2 hours

Paint a room: 3 hours

Complete math homework: 1.5 hours

What do we mean by "work problems"?

What is the concept for solving work problems?

Formula:

EXAMPLE 4 Robert can paint the kitchen in 2 hours. Susan can paint the kitchen in 3 hours. How long will it take Robert and Susan to paint the kitchen if they work together?

	Number of hours to complete the job	Part of the job completed in 1 hour
Robert		
Susan		
Together		

367

EXAMPLE 5 Miguel can rake up the leaves in his yard in 6 hours. His neighbor's son can clean up the leaves in Miguel's yard using his leaf blower. If it takes them 1.5 hours to clear the yard together, how long would it take Miguel's neighbor's son to do the job alone?

	Number of hours to complete the job	Part of the job completed in 1 hour
Miguel		
Neighbor's son		
Together		

EXAMPLE 5 A rain barrel can be filled from the rain gutter of a house during a heavy rainstorm in 3 hours. During the last storm the barrel sprung a small leak which caused the barrel to empty in 8 hours. Now how long does it take to fill the rain barrel. Use a negative number to represent the water lost from the barrel.

	Number of hours to fill the barrel	Fraction of the barrel filled in 1 hour
Gutter		
Leak		
Together		

CONCEPT CHECK: A heater will raise the indoor temperature and an open window will lower the temperature. Which of these rates will involve a negative number?

GUIDED EXAMPLES:

1. If Sabrina can plow the driveway in 4 hours, how much of the task can she complete in an hour?

2. Alfred can mow the huge lawn at his farm with his new lawn mower in 4 hours. His young daughter can mow the lawn with the old lawn mower in 5 hours. If they work together, how long will it take them to mow the lawn?

3. Emmet is using a hose to fill a washtub that has a small hole in the bottom. If the hose can fill the washtub in $\frac{1}{2}$ hour, and the hole will drain the tub in 3 hours, how long will it take Emmet to fill the washtub?

Notebook 28.6
Applications of Rational Equations: $D = RT$

How are distance, rate, and time related?

EXAMPLES 1 & 2 Solve the formula $d = rt$ for each of the variables below.
rate, r

time, t

EXAMPLE 3 The speed of the current in a river is 5 miles per hour. A boat travels 20 miles downstream (with the current) in this river in the same time that it travels 10 miles upstream (against the current). Find the speed of the boat in still water?

	Distance	Rate	Time
Upstream			
Downstream			

EXAMPLE 4 At the airport, Mia and Eric are walking at the same speed to catch their flight, but Eric decides to walk on the moving sidewalk, while Mia continues to walk on the stationary sidewalk. If the sidewalk moves at 1 meter per second, and it takes Eric 50 seconds less to walk the 300-meter distance, at what speed are Mia and Eric walking?

	Distance	Rate	Time
Mia			
Eric			

CONCEPT CHECK: Jamaal drove 45 miles per hour on a trip. If he had driven 60 miles per hours, he would have arrived at his destination 2 hours earlier. Write the equation can be used to find the distance that he drove.

GUIDED EXAMPLES:

1. The speed of the current in a river is 3 miles per hour. A boat travels 60 miles upstream in the same time that it travels 100 miles downstream. Find the speed of the boat in still water.

2. Mandy drove 51 miles per hour on a trip. If she had driven 60 miles per hour, she would have arrived 3 hours earlier. What is the distance that she drove?

370

Notebook 29.1
Direct Variation

Suppose that you earn $8 per hour at a part-time job.
The amount of money you earn depends on the number of hours
that you work.

Hours Worked	0	1	2	3	4
Money Earned					

As the number of hours worked _____, the earnings _____ as
well.

What is direct variation?

Write the equation of direct variation.

What is the constant of variation (ie, the constant of proportionality)?

EXAMPLES 1 & 2 Suppose y varies directly as x. Find the constant of
variation and the equation of direct variation for the following.
$y = 12$ when $x = 3$

$y = -10$ when $x = -2$

EXAMPLES 3 & 4 Find the variation equation in which y varies directly as x
and the following are true. Then find y for the given value of x.
If $y = -6$ when $x = -3$, find y when $x = 10$.

If $y = 15$ when x = 20, find y when $x = 8$.

371

Note: What type of equation is the equation of direct variation?

Properties of Direct Variation:
1. As x _____, y _____
2. As x _____, y _____
3. The graph of $y = kx$ is a _____ line
4. The line always passes through _____
5. The slope of the graph $y = kx$ is _____
6. The equation $y = kx$ is a _____
7. The domain is _____
   and the range is _____

CONCEPT CHECK:
Is this an example of direct variation?
As more food is purchased, the amount of sales tax paid increases.

GUIDED EXAMPLES:
1. Suppose y varies directly as x. If $y = 21$ when $x = 7$, find the constant of variation and the direct variation equation.

2. Suppose y varies directly as x. If $y = -18$ when $x = -3$, find the constant of variation and the direct variation equation.

3. Find the variation equation in which y varies directly as x and $y = -20$ when $x = -4$. Then find y when $x = -6$.

4. Find the variation equation in which b varies directly as a and $b = 8$ when $a = 12$. Then find b when $a = 9$.

Notebook 29.2
Inverse Variation

OTHER NOTES

Suppose that you need to drive a distance of 100 miles.
The faster you drive, the sooner you will arrive at your destination.

Speed (mph)	10	20	25	50
Money Earned				

As the speed _____, the time _____.

What is inverse variation?

Write the equation of inverse variation.

What is the constant of variation (ie, the constant of proportionality)?

EXAMPLES 1 & 2 Find an equation of variation in which y varies inversely
as x
for each of the following.
$y = 5$ when $x = 2$

$y = -4$ when $x = -7$

EXAMPLES 3 & 4 Find the equation of variation in which y varies inversely
as x and the following are true. Then find y for the given value of x.
If $y = -8$ when $x = -7$, find y when $x = 4$.

If $y = 18$ when $x = \dfrac{1}{2}$, find y when $x = 27$.

Note: What type of equation is the inverse variation equation?

373

Properties of Inverse Variation: Sketch the graph here:

1. As x _____, y _____

2. As x _____, y _____

3. The graph of $y = \dfrac{k}{x}$ is _____ a straight line

4. The equation $y = \dfrac{k}{x}$ describes a _____

7. The domain is _____

   and the range is _____

CONCEPT CHECK:

Is this an example of inverse variation?

As the number of dogs increase in a household, the amount of dog food needed goes up.

GUIDED EXAMPLES:

1. Find an equation of variation in which y varies inversely as x and $y = 4$ when $x = 11$.

2. Find an equation of variation in which y varies inversely as x and $y = -9$ when $x = -1$.

3. Find the variation equation in which y varies inversely as x and $y = -12$ when $x = -3$. Then find y when $x = -4$.

4. Find the variation equation in which y varies inversely as x and $y = 10$ when $x = 0.2$. Then find y when $x = -1$.

Name: _____   Date: _____

Instructor: _____   Section: _____

OTHER NOTES

If y varies directly as x, or y is directly proportional to x, then there is a positive constant k such that _____.
The equation $y = kx$ is an _____.

Example.

If y varies inversely as x, or if y is inversely proportional to x,
then there is a positive constant k such that _____.
The equation _____ is called an _____.

Example.

What is joint variation?
A variable can have _____ to two or more variables at the same time. When a variable is _____ to the product of two or more variables, we have _____.

If y varies _____ as x and z or if y is _____ to x and to z,
then there is a _____ k such that _____.

Write the equation of inverse variation.

What is the constant of variation?

EXAMPLE 1 Suppose y varies jointly with x and z. When $x = 2$ and $z = 9$, then $y = 54$.
Find the constant of variation and the equation of joint variation.

EXAMPLE 2 Suppose y varies jointly with x and the square of z. When $x = 3$ and $z = 2$, then $y = 10$. Find y if $x = 5$ and $z = 6$.

What is combined variation?
A variable can have _____ and _____
to two or more variables at the same time. When _____
and _____ happen at the same time, we have
_____.

375

If y varies _____ as x and _____ as z, or if y is
_____ to x and _____ to z,
then there is a _____ k such that _____.

Write the equation of combined variation.
What is the constant of variation?

Note: If a variable has both direct and inverse variation, which variation
is written in the numerator and which is written in the denominator?

EXAMPLE 3 Suppose y varies directly with x and inversely with z.
When $x = 6$ and $z = 9$, then $y = 8$. Find the constant of variation and the
equation of combined variation.

EXAMPLE 4 Suppose y varies directly with the square of x and inversely with
z. When $x = 5$ and $z = 4$, then $y = 12\frac{1}{2}$. Find y if $x = 1$ and $z = 7$.

EXAMPLE 5 Suppose y varies directly with a and the square of b and
inversely with the square of c. When $a = 2$, $b = 3$, and $c = 4$, then $y = \frac{63}{8}$.
Find y if $a = 3$, $b = 1$, and $c = 7$.

CONCEPT CHECK: Is $y = kxz^2$ an equation of joint variation?

GUIDED EXAMPLES:

1. Suppose y varies jointly with x or z. When $x = 3$ and $z = 2$, then $y = 8$. Find
   the constant of variation and the equation of joint variation.

2. Suppose y varies directly with x and inversely with z. When $x = 12$ and
   $z = 3$, then $y = 16$. Find the constant of variation and the equation of
   combined variation.

3. Suppose y varies directly with the square root of x and inversely with z.
   When $x = 36$ and $z = 5$, then $y = 3.6$. Find y if $x = 49$ and $z = 8$.

4. Suppose y varies directly with the square of a and the square root of b, and
   inversely with c. When $a = 3$, $b = 36$ and $c = 18$, then $y = 48$. Find y if
   $a = 1$, $b = 4$, and $c = 11$.

376

Notebook 29.4
Applications of Variation

If y varies directly as x, or y is directly proportional to x, then there is a positive constant k such that _____.

If y varies inversely as x, or if y is inversely proportional to x, then there is a positive constant k such that _____.

EXAMPLE 1 The distance d that a spring stretches varies directly with the weight w of the object hung on the spring. If a 10-pound fish stretches the spring in a fishing scale 6 cm, how far will a 35-pound fish stretch this spring?

EXAMPLE 2 The time t it takes for a job to be done is inversely proportional to the number of people n who are doing the job. If it takes 4 hours for 15 people to do a job, how long will it take 20 people to do the same job? Assume that the people work at the same rate.

EXAMPLE 3 If the voltage in an electric circuit is kept at the same level, the current I varies inversely with the resistance R. The current measures 40 amperes when the resistance is 300 ohms. Find the current when the resistance is 100 ohms.

If y varies directly as x and z, or if y is directly proportional to x and to z, then there is a positive constant k such that _____.

If y varies directly to x and inversely as z, or y is directly proportional to x and inversely proportional to z, then there is a positive constant k such that _____.

Note: If a variable has both direct and inverse variation, the direct variation is written in the _____ and the inverse variation in the _____.

EXAMPLE 4 The force F of the wind on a flat surface placed at a right angle to the direction of the wind varies jointly with the area of the surface A and the square of the speed s of the wind. The force of a 30-miles per hour wind on a window with an area of 20 square feet has a force of 150 pounds. What is the force of a Category 4 hurricane with a wind speed of 130 miles per hour on a window that measures 3 feet by 4 feet?

EXAMPLE 5 The load capacity U of a cylindrical concrete column varies directly with the diameter d of the column raised to the fourth power, and inversely with the square of the length l of the column. If the diameter of a column is 1 foot, and the length is 10 feet, the load capacity is 184 pounds. Find the load capacity if the diameter is 1.5 feet and the length is 12 feet.

CONCEPT CHECK: A child's allowance a increases proportionally based on the number of chores c he completes. What kind of variation does this represent?

GUIDED EXAMPLES:

1. The amount of money you earn S varies directly with the number of hours t that you work. If you earn $60 when you work 5 hours, determine how much money you will earn if you work 8 hours.

2. The cost C of producing a certain pen varies inversely as the number produced n. If 10,000 boxes of pens are produced, the cost is $2 per box. Find the cost per box to produce 25,000 boxes of pens.

3. The force F on the blade of a wind generator varies jointly with the product of the blade's area a and the square of the wind speed s. The force of the wind is 20 pounds when the area is 3 square feet and the wind speed is 30 feet per second. Find the force when the area is increased to 5 square feet and the wind speed is reduced to 25 feet per second. If necessary, round to the nearest hundredth.

4. The electrical resistance of a wire R varies directly with the length of the wire l and inversely with the square of the diameter of the wire d. If a wire 432 feet long and 4 mm in diameter has a resistance of 1.24 ohms, find the resistance in a wire that is 282 feet long with a diameter of 3 mm.

Name: _____   Date: _____

Instructor: _____   Section: _____

Notebook 30.1
Square Roots

The _____ is one of two identical factors of a number.
What does it mean to "take the square root" of a number?

What symbol is used to denote the positive square root of a number?

How do we denote the negative square root?

EXAMPLES 1–4 Find the square roots.

$\sqrt{100}$

$\sqrt{\dfrac{9}{25}}$

$-\sqrt{49}$

$\sqrt{0}$

EXAMPLES 5 & 6 Simplify.

$\sqrt{5^2}$

$\sqrt{(-5)^2}$

For any real number a, $\sqrt{a^2}$ = _____.

We are going to assume that if a variable appears in the radicand, then
it represents a …

EXAMPLES 7 & 8 Find the square roots.

$\sqrt{49x^2}$

$\sqrt{(-6x)^2}$

What happens when you try to simplify $\sqrt{-25}$?

What can you say about the square root of any negative number?

What about a negative square root of a negative number, like $-\sqrt{-36}$?
Explain.

379

EXAMPLES 9 & 10 Find the square roots.

$\sqrt{-144}$

$-\sqrt{-9}$

EXAMPLE 11 Use a calculator to approximate $\sqrt{3}$.
Round to the nearest hundredth.

Explain how you can approximate a square root if you do not have a calculator.

EXAMPLE 12 Approximate $\sqrt{75}$ by finding two consecutive whole numbers that the square root lies between.

CONCEPT CHECK:
What is the principal square root of 9?

GUIDED EXAMPLES:

1. Find the square root. If the square root is not a real number, say so. $\sqrt{121}$

2. Find the square root. If the square root is not a real number, say so.
   $-\sqrt{-81}$

3. Find the square root. If the square root is not a real number, say so.
   Assume all variables represent nonnegative numbers. $-\sqrt{121a^2}$

4. Use a calculator to approximate the square root to the nearest hundredth.
   $-\sqrt{84}$

380

Name: _____   Date: _____

Instructor: _____   Section: _____

Notebook 30.2
Higher Order Roots

The _____ is one of two identical factors of a number.
What does it mean to "take the square root" of a number?

How is a higher-order root written?
What is the n called?
What is the a called?
Note: What is the index of a square root?

What is the cube root of a number?
What is the index of a cube root?

Example $\sqrt[3]{8}$

Let's look at higher powers...
Complete the table. Keep it handy, you will use it often.

x	x^2	x^3	x^4	x^5
1				
2				
3				
4				
5				
6				
7				
8				
9				
10				

EXAMPLES 1 – 3 Find the cube roots.

$\sqrt[3]{27x^3}$

$-\sqrt[3]{\dfrac{8}{27}}$

$\sqrt[3]{-1}$

Why is it ok to have a cube root of a negative number?
The cube root of a negative number is a _____.

What is meant by the nth root?

EXAMPLES 4 & 5 Find the indicated roots.

$\sqrt[5]{32}$

$\sqrt[4]{-81}$

When must the radicand be nonnegative?
It is impossible to take an _____ root of a _____ number!

EXAMPLES 6–8 Simplify the radicals.

$\sqrt[5]{x^{10}}$

How can you find an nth root when the radicand contains variables?

$\sqrt[4]{y^{12}}$

$\sqrt[4]{a^{24}b^{28}}$

EXAMPLES 9 & 10 Simplify the radicals.

$\sqrt[3]{-64x^6}$

$\sqrt[4]{10,000x^4}$

CONCEPT CHECK:
Write a radical expression that has an index of 4.

GUIDED EXAMPLES:

1. Find the cube root. $\sqrt[3]{-64}$

2. Find the cube root. $-\sqrt[3]{\dfrac{a^6}{125}}$

3. Find the indicated root. $\sqrt[5]{x^{30}}$

4. Find the indicated root. $\sqrt[5]{-32x^5}$

Notebook 30.3
Simplifying Radical Expressions

The _____ is one of two identical factors of a number.
Examples.

What does the Product Rule for Radicals state?

Again, what assumption will be made about variables inside radicals?

EXAMPLE 1 Simplify the radical. $\sqrt{50}$

EXAMPLES 2 & 3 Find the indicated roots.
$\sqrt{8}$

$\sqrt{48}$

EXAMPLE 4 Simplify, if possible. $\sqrt[3]{-54}$

How do you find an nth root?
Example.

EXAMPLES 5 & 6 Simplify the radicals. Assume all variables represent nonnegative values.
$\sqrt{x^5}$

$\sqrt[3]{x^{17}}$

EXAMPLES 7 & 8 Simplify the radicals. Assume all variables represent nonnegative values.
$\sqrt{24x^2}$

$\sqrt[3]{-16x^{11}}$

How can you simplify radicals if you do not know perfect root factors?

EXAMPLE 9 Simplify. Assume all variables represent nonnegative values.
$$\sqrt[3]{500x^5y^7}$$

CONCEPT CHECK:

What is the remainder when $\sqrt[6]{x^{17}}$ is simplified?
(That is, what remains under the radical?)

GUIDED EXAMPLES:

1. Simplify. Assume all variables represent nonnegative values. $\sqrt{1000a^3}$

2. Simplify. $\sqrt[3]{250y^3}$

3. Simplify. $\sqrt[3]{\dfrac{81}{8}}$

4. Simplify. $\sqrt[3]{128x^5y^4}$

Notebook 30.4
Rational Exponents

How do you write rational exponents as radicals?

If n is an integer greater than 1, with $\dfrac{1}{n}$ in simplest form,

then…

The denominator of the rational exponent becomes the _____.

EXAMPLES 1 & 2 Write in radical notation. Simplify, if possible.

$9^{\frac{1}{2}}$

$x^{\frac{1}{3}}$

Note: What does a $\dfrac{1}{2}$ power mean?

What does a $\dfrac{1}{3}$ power mean?

What is the rule for writing rational exponents as radicals?

If m and n are integers greater than 1, with $\dfrac{m}{n}$ in simplest form,

then…

EXAMPLES 3 & 4 Write in radical notation. Simplify, if possible.

$6^{\frac{2}{3}}$

$x^{\frac{3}{2}}$

EXAMPLES 5 & 6 Use radical notation to rewrite each expression.
Simplify, if possible.

$\left(\dfrac{1}{9}\right)^{\frac{3}{2}}$

$(3x)^{\frac{1}{3}}$

What does a negative exponent mean?
How do you simplify an exponential expression that has a negative
rational exponent?

EXAMPLES 7 & 8 Evaluate the expressions.

$25^{-\frac{3}{2}}$

$-16^{-\frac{3}{4}}$

Write the product rule for exponents.
Write the quotient rule for exponents.

EXAMPLES 9 & 10 Use the rules of exponents to simplify.

$\dfrac{6^{\frac{1}{3}}}{6^{\frac{4}{3}}}$

$x^{\frac{1}{2}} \cdot x^{\frac{1}{3}}$

CONCEPT CHECK:

What would be the fastest way to simplify $\dfrac{7^{\frac{5}{6}}}{7^{\frac{2}{3}}}$?

GUIDED EXAMPLES:

1. Write in radical form. Simplify if possible. Assume all variables represent nonnegative values. $-x^{\frac{3}{5}}$

2. Simplify. $9^{\frac{5}{2}}$

3. Simplify. Write your answer in radical notation. Assume all variables represent nonnegative values. $x^{\frac{2}{5}} \cdot x^{\frac{1}{2}}$

4. Simplify. $\dfrac{7^{\frac{1}{5}}}{7^{\frac{6}{5}}}$

Notebook 30.5
More on The Pythagorean Theorem

A _____ is an angle that measures 90°.
A triangle that has one right angle is a _____.

The sides that form the right angle are called the _____.
The side opposite the _____ is the hypotenuse.
The _____ is the longest side of a right triangle.
Label the sides and right angle of the triangle.

State the Pythagorean Theorem.
In a _____, the sum of the _____ of the legs
is equal to the square of the _____, or …

Substituting a and b for the _____ and c as the _____,
what does the Pythagorean Theorem become?

EXAMPLE 1 Find the length of the hypotenuse. Label each side.

Identify a, b, and c.
Substitute values into Pythagorean Theorem.
Solve.

EXAMPLE 2 Find the length of the missing leg. Label each side.

Identify a, b, and c.
Substitute values into Pythagorean Theorem.
Solve.

EXAMPLE 3 Find the length of the hypotenuse. Label each side.

Identify a, b, and c.
Substitute values into Pythagorean Theorem.
Solve.

EXAMPLE 4 A slanted roof rises 5 feet vertically from the edge of the roof to the top. The roof covers 12 horizontal feet. How long is the slanted surface of the roof?

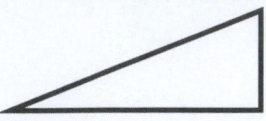

Identify a, b, and c.
Substitute values into Pythagorean Theorem.
Solve.

CONCEPT CHECK:
True/False The hypotenuse is the longest side of a right triangle.

GUIDED EXAMPLES
1. Find the length of the hypotenuse.

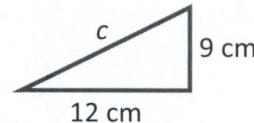

2. Find the length of the missing leg.

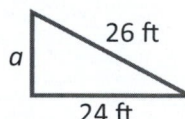

3. Find the length of the missing leg. If necessary, round to the nearest hundredth.

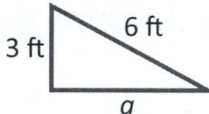

4. A slanted roof rises 24 feet vertically from the edge of the roof to the top. The roof covers 7 horizontal feet. How long is the slanted surface of the roof?

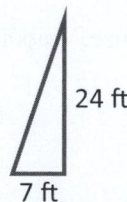

388

Notebook 30.6
The Distance Formula

The _____ is a way to find the missing
side of a right triangle.

To find the hypotenuse of a right triangle, what formula do you use?

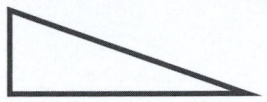

EXAMPLE 1 Find the length of the line segments between the two points
shown.

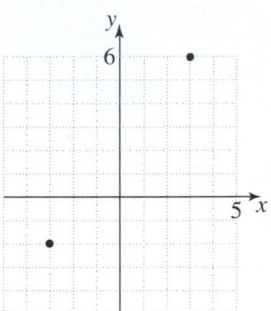

Note. How can you find the horizontal distance between two points?
How can you find the vertical distance between two points?

EXAMPLE 2 Find the horizontal distance and the vertical distance between
the points (–3, 2) and (1, 5).

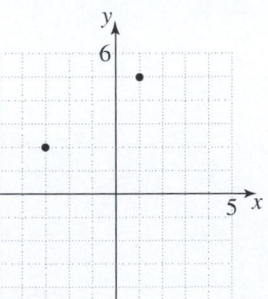

EXAMPLE 3 Find the distance between (–3, 2) and (1, 5).   (Use graph
above.)

389

The _____ is used to calculate the distance between any two points. Write the Distance Formula here:

EXAMPLES 4 & 5 Use the Distance Formula to find the distance between the given points. If necessary, round to the nearest hundredth.
(6, –3) and (1, 9)

(2, 1) and (6, 8)

EXAMPLE 6 Use the Distance Formula to find the distance between the given points. If necessary, round to the nearest hundredth.
(–13, –8) and (–3, –3)

CONCEPT CHECK:
How is the horizontal distance between two points calculated?

GUIDED EXAMPLES:
1. Find the distance between the two points shown on the graph.

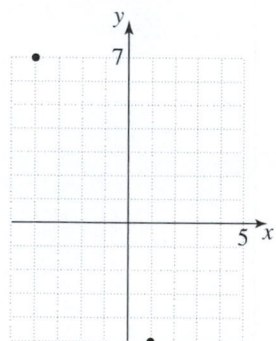

2. Find the distance between (2 –7) and (–1, –11).

3. Find the distance between (–6, 2) and (–1,6). If necessary, round to the nearest hundredth.

4. Find the distance between (–4, –10) and (–1, –4). If necessary, round to the nearest hundredth.

390

Name: _____   Date: _____

Instructor: _____   Section: _____

Notebook 31.1
Introduction to Radical Functions

How do you evaluate a function at a certain value?

EXAMPLES 1–3 If $f(x) = \sqrt{2x+4}$, find the function values.

$f(2.5)$

$f(-3)$

$f(-2)$

What is the domain of a function?

The square root of a negative number is _____.

How do you find the domain of a function?

EXAMPLE 4 Find the domain of $f(x) = \sqrt{x+2}$.

EXAMPLE 5 Graph $f(x) = \sqrt{x+2}$.

x	$f(x)$
-2	
-1	
0	
1	
2	
7	

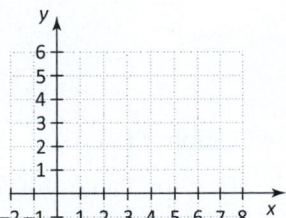

EXAMPLE 6 Graph $f(x) = \sqrt{3x-9}$.

x	$f(x)$
3	
4	
5	
6	

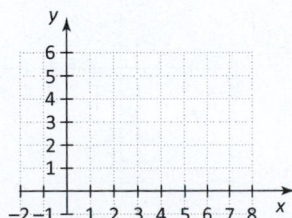

391

EXAMPLE 7 Graph $f(x) = \sqrt[3]{x}$.

x	$f(x)$
0	
1	
−1	
6	
−6	
8	
−8	

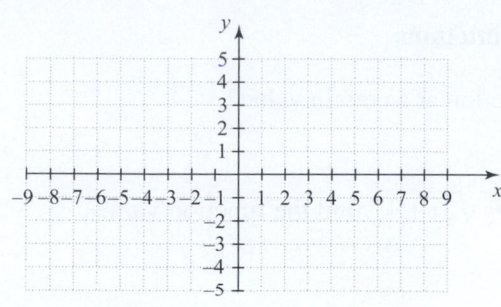

EXAMPLE 8 The driver of a car suddenly slams on the brakes and skids on wet pavement. The speed of the car, S, in miles per hour can be approximated by the function $s(l) = 2\sqrt{3l}$, where l is the length of the side marks in feet. If the skid marks on the road are 17 feet long, how fast was the car going? Round your answer to the nearest whole number if necessary.

Understand the problem

Create a plan

Find the answer

Check the answer

CONCEPT CHECK:

If $f(x) = \sqrt{x+2}$, is $f(-3)$ a real number?

GUIDED EXAMPLES:

1. If $f(x) = \sqrt{7x-20}$, find $f(10)$.

2. Find the domain of $\sqrt{x-2}$.

3. Graph $f(x) = \sqrt{x-2}$. Use the function values for $f(2), f(3), f(4), f(6)$, and $f(8)$.

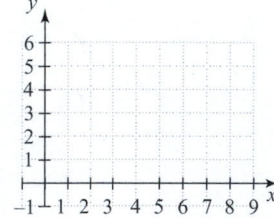

4. Graph $f(x) = \sqrt[3]{x-1}$. Use the function values for $f(1), f(0), f(-1), f(2), f(-2), f(3)$, and $f(-3)$.

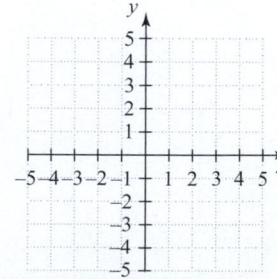

392

Name: _____   Date: _____

Instructor: _____   Section: _____

Notebook 31.2
Adding and Subtracting Radical Expressions

OTHER NOTES

_____ are terms that have the same variables raised to the same powers.

EXAMPLES 1–3 Combine like terms.

$7x + 9x$

$13xy^2 + 11x^2y$

$5x^{1/2} + 2x^{1/2}$

What does the rational rule for exponents say?

If m and n are real numbers greater than 1 with $\dfrac{m}{n}$ in simplest form, then…

A _____ is written using the radical sign, _____, where n is the _____ and a is the _____.

What are like radicals?

Give some examples of like radicals.

Give some examples of unlike radicals. Tell why they are unlike.

Write the steps for adding and subtracting radical expressions.
1.
2.
Note: Only _____ can be added or subtracted.

EXAMPLES 4–6 Simplify, if possible. Assume that all variables are nonnegative real numbers.

$5\sqrt{3} + 7\sqrt{3}$

$3\sqrt{2} - 9\sqrt{2}$

$-6\sqrt{xy} + 2\sqrt[3]{xy}$

The Product Rules for radicals says that for all nonnegative real real numbers a, b, and n, where n is an integer greater than 1, …

393

EXAMPLE 7 Simplify. $\sqrt{50}+\sqrt{18}$

Simplify each radical term.

Combine like radicals.

EXAMPLE 8 Simplify. $5\sqrt{3}-\sqrt{27}+2\sqrt{32}$

Simplify each radical term.

Combine like radicals.

EXAMPLE 9 Simplify. Assume that all variables are nonnegative real numbers. $6\sqrt{x}+4\sqrt{12x}-\sqrt{75x}+3\sqrt{x}$

EXAMPLE 10 Simplify. Assume that all variables are nonnegative real numbers. $3x\sqrt[3]{54x^4}-3\sqrt[3]{16x^7}$

CONCEPT CHECK:

Can $\sqrt{50}$ and $3\sqrt{5}$ be combined?

GUIDED EXAMPLES:

1. Simplify. If the radicals cannot be combined, state why. Assume variables represent nonnegative numbers.

   $4\sqrt{11}-7\sqrt{11}$

2. Simplify. $\sqrt{54}+\sqrt{24}$

3. Simplify. Assume variables represent nonnegative numbers.

   $\sqrt{18x}-x\sqrt{80x}+7\sqrt{2x}-\sqrt{5x^3}$

4. Simplify. $\sqrt[3]{24n^7}-4n\sqrt[3]{81n^4}$

Name: _____   Date: _____

Instructor: _____   Section: _____

Notebook 31.3
Multiplying Radical Expressions

The Distributive Property says that for a, b, and c, $a(b + c) =$ _____.

EXAMPLE 1 Multiply.

$-6\left(\sqrt{2} + \sqrt{3}\right)$

The Product Rule for Radicals states that for all nonnegative real numbers a, b, and n, where n is an integer greater than 1, …

EXAMPLE 2 Multiply.

$\sqrt{5} + \sqrt{3}$

To multiply monomials, first multiply the _____, and then _____ the variables.

Multiply. $4x \cdot 3y$

EXAMPLE 3 Multiply. Assume all variables represent nonnegative numbers.

$-7x \cdot 9xy$ $-7\sqrt{x} \cdot 9\sqrt{2x}$

EXAMPLE 4 Multiply. $\left(\sqrt{12}\right)\left(-5\sqrt{3}\right)$

EXAMPLE 5 Multiply. $\left(\sqrt[3]{2x}\right)\left(\sqrt[3]{4x^2} + 3\sqrt[3]{y}\right)$

To multiply binomials,
1.
2.

Multiply $(a + b)(c + d)$ using the FOIL method.

EXAMPLE 6 Multiply. $\left(\sqrt{2}+3\sqrt{5}\right)\left(2\sqrt{2}+\sqrt{5}\right)$

Note: Remember, $\left(\sqrt{x}\right)^2 =$

Examples.

$\left(\sqrt{3}\right)^2$ $\left(\sqrt{7y}\right)^2$

If a and b are numbers or expressions, $(a+b)(a-b) =$

EXAMPLE 7 Multiply. $\left(5+3\sqrt{2}\right)\left(5-3\sqrt{2}\right)$

If a and b are numbers or expressions, $(a+b)^2 =$
and $(a-b)^2 =$

EXAMPLE 8 Simplify. Assume all variables represent nonnegative numbers.
$\left(\sqrt{7}+\sqrt{3x}\right)^2$

CONCEPT CHECK:

How many radicals will be left in the final answer when $\left(\sqrt{2}-3\sqrt{6}\right)\left(\sqrt{2}-\sqrt{6}\right)$
is multiplied?

GUIDED EXAMPLES:

1. Multiply. $\left(3\sqrt{28}\right)\left(4\sqrt{2}\right)$

2. Multiply. Assume all variables represent nonnegative numbers.
$\sqrt{6x}\left(2\sqrt{3x}+5\sqrt{2}\right)$

3. Multiply. $\left(6+2\sqrt{5}\right)\left(8-\sqrt{7}\right)$

4. Simplify. Assume all variables represent nonnegative numbers.
$\left(\sqrt{3x}-2\sqrt{5}\right)^2$

Name: _____   Date: _____

Instructor: _____   Section: _____

Notebook 31.4
Dividing Radical Expressions

For all nonnegative real numbers a, b, and n, where n is an integer greater than 1, $\sqrt[n]{a} \cdot \sqrt[n]{b} =$ _____ (The Product Rule for Radicals)

The Quotient Rule for Radicals states that for all nonnegative real numbers a, b, and n, where n is an integer greater than 1 and $b \neq 0$, $\dfrac{\sqrt[n]{a}}{\sqrt[n]{b}} =$

EXAMPLE 1 Divide. $\dfrac{\sqrt{75}}{\sqrt{3}}$

Write the procedure for dividing radical expressions.
If you can easily take the _____ of both the _____
and _____, take the _____ first, then _____.
If you can easily _____ first, _____, and then
take the _____.

EXAMPLES 2 & 3 Divide.

$\dfrac{\sqrt{9}}{\sqrt{16}}$

$\dfrac{\sqrt{72}}{\sqrt{8}}$

EXAMPLES 4 & 5 Divide. Assume all variables represent nonnegative values.

$\sqrt[3]{\dfrac{-a^9}{b^6}}$

$\dfrac{\sqrt[3]{x^8}}{\sqrt[3]{x^5}}$

397

EXAMPLE 6 Divide. Assume all variables represent nonnegative values.

$$\frac{\sqrt{54a^3b^7}}{\sqrt{6b^5}}$$

EXAMPLE 7 Divide. Assume all variables represent nonnegative values.

$$\frac{\sqrt{25x^5y^2}}{\sqrt{144x^3y^4}}$$

CONCEPT CHECK:

For $\dfrac{\sqrt{72}}{\sqrt{8}}$ would it be easier to take the root of the fraction or divide first and then take the root?

GUIDED EXAMPLES:

1. Divide. Assume all variables represent nonnegative numbers.

$$\frac{\sqrt{98}}{\sqrt{2}}$$

2. Divide. $\dfrac{\sqrt{144}}{\sqrt{4}}$

3. Divide. Assume variables represent positive values.

$$\frac{\sqrt{48a^9b^5}}{\sqrt{3a^7b}}$$

4. Divide. Assume variables represent positive values.

$$\sqrt{\frac{80a^5b^3}{5a^3b^9}}$$

398

Name: _____   Date: _____

Instructor: _____   Section: _____

Notebook 31.5
Rationalizing the Denominator

A _____ can be written as a fraction of two integers.
A rational number written in decimal form will either _____
or _____.
Write some examples of rational numbers.

What are irrational numbers?
Give some examples.

What does the Identity Property of Multiplication say?

Rationalizing the Denominator means…
removing a _____ from a denominator.
This means changing the _____ from a _____
number to a _____ number.

How do you "rationalize a denominator"?

EXAMPLE 1 Simplify by rationalizing the denominator. $\dfrac{1}{\sqrt{2}}$

EXAMPLE 2 Simplify by rationalizing the denominator. $\dfrac{3x}{\sqrt{12x^2}}$

EXAMPLE 3 Simplify by rationalizing the denominator. $\sqrt[3]{\dfrac{2}{3x^2}}$

Remember, $(a + b)(a - b) =$

399

What are conjugates?

Give some examples of conjugates. What are their products?

Note. What can you say about the product of two conjugates?

EXAMPLE 4 Simplify by rationalizing the denominator. $\dfrac{5}{3+\sqrt{2}}$

EXAMPLE 5 Simplify by rationalizing the denominator. $\dfrac{\sqrt{7}+\sqrt{3}}{\sqrt{7}-\sqrt{3}}$

CONCEPT CHECK:

What should $\dfrac{2}{3+\sqrt{5}}$ be multiplied by in order to rationalize the denominator?

GUIDED EXAMPLES:

1. Simplify by rationalizing the denominator. Assume that all variables
    represent nonnegative numbers and $y \neq 0$.

    $\dfrac{\sqrt{7x}}{\sqrt{14y}}$

2. Simplify by rationalizing the denominator. $\sqrt[3]{\dfrac{3}{5x^2}}$

3. Simplify by rationalizing the denominator. $\dfrac{4}{2+\sqrt{5}}$

4. Simplify by rationalizing the denominator. $\dfrac{\sqrt{3}+\sqrt{10}}{\sqrt{3}-\sqrt{10}}$

Notebook 31.6
Solving Radical Equations

What are examples of reverse operations?

What does the squaring property of equality say?

Examples.

Write the steps for solving equations containing radicals.
1.

2.
     *

     *

3.

4.

5.

EXAMPLE 1 Solve. $\sqrt{x} = 7$.
Square both sides.

EXAMPLE 2 Solve. $\sqrt{x} = -8$.
Square both sides.

Check.

EXAMPLE 3 Solve. $3 + \sqrt{x-5} = 15$.
Get the radical by itself.

Square both sides.

Solve.

Check.

OTHER NOTES

EXAMPLE 4 Solve. $\sqrt{5x+6} = x$.

Square both sides.

Solve.

Check.

Why is -1 not a solution?

EXAMPLE 5 Solve. $-6 + \sqrt[3]{x} = -10$.

Get the radical by itself.

Cube both sides.

Solve.

Check.

EXAMPLE 6 The Body Mass Index, BMI, is an indirect measure of a person's body fat based on height and weight. One formula involving BMI is

$h = \sqrt{\dfrac{w}{\text{BMI}}}$, where w is the person's weight in kilograms and h is the height in

meters. Solve this equation for BMI, and then use this result to determine if the BMI for a person who is 1.8 meters tall and weighs 66 kilograms is healthy. Note: The healthy BMI range is between 18.5 and 24.9.

CONCEPT CHECK:
True/False Raising a number to a power and taking a root of a power are reverse operations.

GUIDED EXAMPLES:

1. Solve. $\sqrt{x} = 13$

2. Solve. $-6 + \sqrt{x} = 3$

3. Solve. $\sqrt{x+2} = x$

4. Solve. $\sqrt{(x+1)} + 3 = 7$

Name: _____   Date: _____

Instructor: _____   Section: _____

Notebook 32.1
Introduction to Solving Quadratic Equations

What is the standard form of a quadratic equation?
What is the highest degree of any term in a quadratic equation?

EXAMPLES 1–3 Write each quadratic equation in standard form.

$7x^2 = 5x + 8$

$6x - 2x^2 = -3$

$4x^2 = 9$

What is the graph of a quadratic equation called?
What do they look like? (Sketch both cases.)

What is the y-coordinate of the x-intercept?
What is the x-coordinate of the x-intercept?

The _____ of an equation is the number(s) that, when substituted for the variable(s), make(s) the equation true.

The _____ is the graph of an equation is/are the _____ where the curve crosses the _____.

Note: How many x-intercepts will the graph of a quadratic equation have?

The _____ to the quadratic equation are the _____ of the x-intercepts.

EXAMPLES 4–6 Determine the number of solutions for each quadratic graphed below.

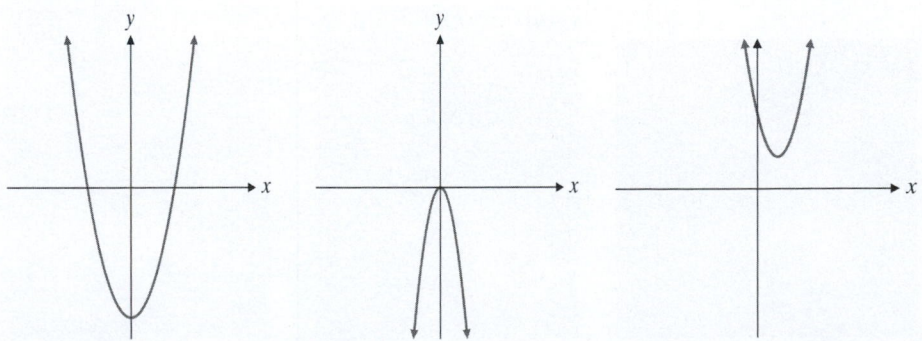

403

EXAMPLES 7 & 8 Find the real solution(s) of each quadratic equation, if any exist, by using the graph of the equation.

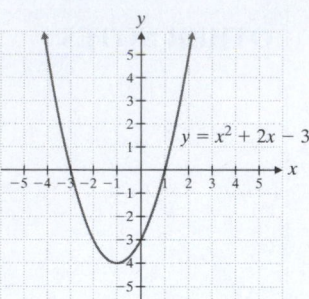

$y = x^2 + 2x - 3$

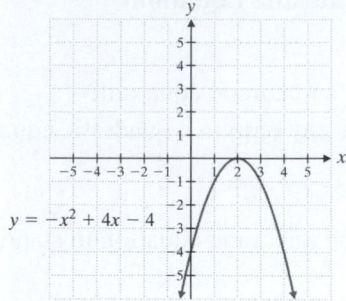

$y = -x^2 + 4x - 4$

CONCEPT CHECK:

If a quadratic equation has two solutions, how many x-intercepts does it have?

GUIDED EXAMPLES: See guided examples.

1. Write each equation in standard form. $8x + 1 = 5x^2$

2. Determine how many solutions the quadratic equation represented by each graph has.

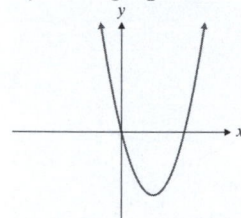

3. Find the real solution(s) of the quadratic equation, if any exist, by using the graph of the equation.

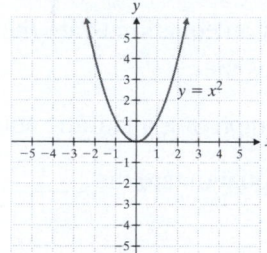

$y = x^2$

4. Find the real solution(s) of the quadratic equation, if any exist, by using the graph of the equation..

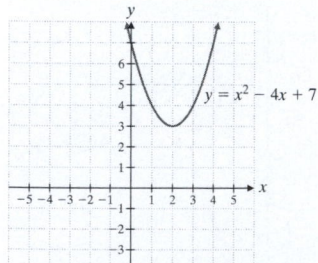

$y = x^2 - 4x + 7$

404

Name: _____   Date: _____

Instructor: _____   Section: _____

Notebook 32.2
Solving Quadratic Equations by Factoring

Write the steps for solving a quadratic equation by factoring.
1. Make sure the equation is in _____.
2. _____, if possible.
3. Set each factor _____.
4. _____ the resulting equations.
5. _____ each solution in the original equation.

Note. What happens if the quadratic expression does not factor?

EXAMPLE 1 Solve by factoring. $x^2 - 9x = 0$
Factor.
Set each factor $= 0$.
Solve.
Check.

EXAMPLE 2 Solve by factoring. $5x^2 - 14x - 3 = 0$
Factor.
Set each factor $= 0$.
Solve.
Check.

EXAMPLE 3 Solve by factoring. $9x^2 = 24x - 15$
Write in standard form (set $= 0$).
Factor.
Set each factor $= 0$.
Solve.
Check.

OTHER NOTES

EXAMPLE 4 Solve by factoring. $x^2 - 2x = -1$

Write in standard form (set = 0).

Factor.

Set each factor = 0.

Solve.

Check.

CONCEPT CHECK:

What is the first step to solve $x^2 = 3x$ by factoring?

GUIDED EXAMPLES:

Solve by factoring.

1. $2x^2 + 16x = 0$

2. $3x^2 + 8x - 35 = 0$

3. $5x^2 + 60 = 35x$

4. $x^2 + 6x + 9 = 0$

Notebook 32.3
Solving Quadratic Equations Using the Square Root Property

A _____ is an equation that can be written in the
form _____ where $a \neq 0$.
This is called _____ form of a quadratic equation.
What is the highest degree of any term in a quadratic equation?

What does the Square Root Property (SRP) say?

Note: What is the shorthand way of writing $+\sqrt{a}$ or $-\sqrt{a}$?

EXAMPLE 1 Solve for x. $x^2 = 36$
Use the SRP.
Simplify.
Check.

Caution! Don't forget to use the _____. Otherwise, you will get
_____ of the two solutions.

EXAMPLE 2 Solve for x. $x^2 = 48$
Use the SRP.
Simplify.
Check.

Caution! The Square Root Property can only be used to solve one type
of equation. What form must it be in?

EXAMPLE 3 Solve for x. $x^2 = -4$
Use the SRP.
Simplify.
Check.

Note: What can you say about an equation of the form $x^2 = a$ negative number?

407

Note: Sometimes you will need to perform steps to get the _____ alone on one side of the equation.

EXAMPLE 4 Solve for x. $3x^2 + 2 = 77$

Note: Can the Square Root Property be used to solve quadratic equations in which the base is an algebraic expression?

EXAMPLE 5 Solve for x. $(4x - 1)^2 = 5$

CONCEPT CHECK:
True/False The Square Root Property can only be used for equations in the form of $x^2 = a$ if a is a perfect square.

GUIDED EXAMPLES:
1. Solve for x.
   $$x^2 = 49$$

2. Solve for x.
   $$x^2 = 18$$

3. Solve for x.
   $$x^2 = -25$$

4. Solve for x.
   $$(2x + 3)^2 = 7$$

Name: _____   Date: _____

Instructor: _____   Section: _____

Notebook 32.4
Solving Quadratic Equations by Completing the Square

The _____(SRP) states that if $x^2 = a$ for $a \geq 0$,
then $x =$ _____.

Examples. $x^2 = 18$ $(x + 2)^2 = 9$ $(5x - 2)^2 = 3$

Is the equation $x^2 + 6x + 9 = 4$ in the correct form to use the SRP?
How can you get it in the correct form?
Solve.

What makes a perfect square trinomial?
The _____ and _____ terms are perfect squares and the middle
term is _____ the product of the numbers being _____ in the first and
last terms.
A perfect square trinomial is of the form $(a + b)^2 =$
or $(a - b)^2 =$

EXAMPLES 1–2 Fill in the blanks to create a perfect square trinomial.
$x^2 + 8x +$ _____ $= (x +$ ____$)^2$

$x^2 - 12x +$ _____ $= (x -$ ____$)^2$

This is called "_____".
Note: To get the constant term of a perfect square trinomial where $a = 1$,
find _____ of the coefficient of ___, and _____ the result.

EXAMPLE 3 Solve $x^2 + 2x = 3$ by filling in the blanks and then using the
Square Root Property.
$x^2 + 2x +$ _____ $= 3 +$ _____

Write the steps for solving a quadratic equation by completing the square.

1. If the coefficient of x^2 is 1, …
   If not, get the coefficient to 1 by _____ each term of the
   equation by the _____ of x^2.
2. Write the _____ so that all terms with _____ are on one
   side of the equation and all _____ are on the other.
3. Find _____ of the _____ of x and then _____
   the result. _____ to both sides of the equation.
4. _____ the resulting perfect square trinomial. _____
   the other side.
5. Use the _____ to solve the resulting equation.

Note: When might an equation be more easily solved by a different method?

EXAMPLE 4 Solve by completing the square. $x^2 + 4x - 5 = 0$

EXAMPLE 5 Solve by completing the square. $x^2 + 6x + 1 = 0$

CONCEPT CHECK:
What is the first step to solving $2x^2 + 12x = -22$ by completing the square?

GUIDED EXAMPLES:
1. Fill in the blanks to create a perfect square trinomial.
   $$x^2 - 10x + \underline{} = (x - \underline{})^2$$

2. Solve by completing the square.
   $$x^2 + 6x - 7 = 0$$

3. Solve by completing the square.
   $$3x^2 + 12x - 36 = 0$$

4. Solve by completing the square.
   $$2x^2 + 4x + 1 = 0$$

Notebook 32.5
Solving Quadratic Equations Using the Quadratic Formula

State the Quadratic Formula. Remember, this is for all quadratic equations of the form $ax^2 + bx + c = 0$.

Write the steps for solving an equation using the Quadratic Formula.
1. Write the equation in _____ form.
   What does that mean?
2. Identify a, b, and c.
3. Substitute a, b, and c, into the _____,
   $$x = \frac{-b \pm \sqrt{b^2 - 4ac}}{2a}.$$
4. _____.

SUMMARY OF SOLVING QUADRATIC EQUATIONS
When might the factoring method be easiest?

When should you consider using the Square Root Property?

When might Completing the Square work nicely?

What is the advantage for using the Quadratic Formula?

EXAMPLE 1 Solve by using the quadratic formula. $3x^2 - x - 2 = 0$
Write in standard form.
Identify a, b, and c.
Substitute.
Simplify.

EXAMPLE 2 Solve by using the quadratic formula. $x^2 = 6x$
Write in standard form.
Identify a, b, and c.
Substitute.
Simplify.

Caution! What will make b or $c = 0$? Is this ok?

EXAMPLE 3 Solve by using the quadratic formula. $4x^2 + 25 = 20x$

Write in standard form.

Identify a, b, and c.

Substitute.

Simplify.

EXAMPLE 4 Solve by using the quadratic formula. $5x^2 - 2x + 3 = 0$

Write in standard form.

Identify a, b, and c.

Substitute.

Simplify.

EXAMPLE 5 Solve by using the quadratic formula. $x^2 + 4x - 8 = 0$

Write in standard form.

Identify a, b, and c.

Substitute.

Simplify.

Caution! Make sure you divide _____ terms in the formula by $2a$.

CONCEPT CHECK:

What are the values of a, b, and c, for the quadratic equation $5x^2 + 2x - 10 = 0$?

GUIDED EXAMPLES:

Solve by using the quadratic formula.

1. $x^2 - 4x + 3 = 0$

2. $x^2 + 5x = -1 + 2x$

3. $2x^2 - 26 = 0$

4. $4x^2 = -4x - 1$

Name: _____   Date: _____

Instructor: _____   Section: _____

Notebook 32.6
Applications with Quadratic Equations

When an object is thrown or launched upward, its approximate height
in meters is given by the quadratic function _____,
where $S(t)$ is the _____, v is the _____,
t represents the _____, and h represents the initial
_____.

For all quadratic equations in the form $ax^2 + bx + c = 0$, $x =$

EXAMPLE 1 A tennis ball is thrown upward with an initial velocity of
10 meters per second. Suppose that the initial height above the ground is
4 meters.

Find the height, S, of the ball after 1 second.

At what time t will the ball hit the ground? Round your answer to the nearest
hundredth. Remember, the function for when an object is thrown or launched
is given by $S(t) = -5t^2 + vt + h$.

EXAMPLE 2 A rocket is fired upward with an initial velocity, v, of 110
meters per second. The quadratic function $S(t) = -5t^2 + 110t$ can be used to find
the heights, S, of the rocket, in meters, at any time t, in seconds.

Find the height of the rocket 8 seconds after it takes off.

Find how many seconds it takes for the rocket to reach a height of 500 meters.
Round your answers to the nearest hundredth.

State the Pythagorean Theorem. In a _____ triangle, the _____ of the
_____ of the _____ is equal to the square of the _____,
or _____ .

413

EXAMPLE 3 Find the lengths of the sides of the triangle. Round your answers to the nearest tenth.

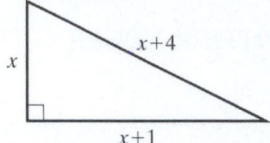

EXAMPLE 4 A backyard pool has a concrete walkway around it so that it is 4 feet wide on all sides. The total area of the pool and the walkway is 1015 ft^2. If the length of the pool is 6feet longer than the width, find the dimensions of the pool. (You will need to sketch the figure in the space below.)

CONCEPT CHECK:
A backyard pool has a concrete walkway around it that is 3 feet wide on all sides. The total area of the pool and the walkway is 996 ft^2. The length of the pool is 5 feet longer than the width. What algebraic expression represents the length of the pool plus the walkway?

GUIDED EXAMPLES:
1. A ball is thrown upward with an initial velocity v of 13 meters per second. Suppose that the initial height h above the ground is 7 meters. At what time t will the ball hit the ground? The ball is on the ground when $S(t) = 0$. Use the function $S(t) = -5t^2 + vt + h$, and round your answer to the nearest hundredth.

2. A rocket is fired upward with a velocity v of 480 feet per second. Find how many seconds it takes for the rocket to reach a height of 3200 feet. Use the quadratic function $S(t) = -16t^2 + vt$.

3. A rectangular sandbox is surrounded on all sides by a strip of grass that is 3 feet wide on all sides. The length of the sandbox is 2 feet more than the width, and the total area for the sandbox and the grass is 528 ft^2. Find the dimensions of the sandbox.

4. One leg of a right triangle is 3 inches more than twice the length of the other leg. The hypotenuse is 9 inches more than the shorter leg. Find the lengths of all the sides. Round your answers to the nearest hundredth.

414

Name: _____   Date: _____

Instructor: _____   Section: _____

Notebook 33.1
Complex Numbers

What is an imaginary number?

What makes up the set of imaginary numbers?

For all positive real numbers a, $\sqrt{-a} =$

EXAMPLES 1 & 2 Simplify.

$\sqrt{-49}$

$\sqrt{-8}$

Imaginary Numbers:

$i =$ $i^5 =$

$i^2 =$ $i^6 =$

$i^3 =$ $i^7 =$

$i^4 =$ $i^8 =$

Note: $i^{4n} =$ _____ where $n \neq 0$.

EXAMPLES 3 & 4 Evaluate the powers of i.

$i^9 =$

$i^{15} =$

What are complex numbers? Numbers of the form _____, where a and b are _____. The a is the _____ and bi is the _____.

For all real numbers a, b, c, and d, $(a + bi) + (c + di) =$ and $(a + bi) - (c + di) =$

EXAMPLES 5 & 6 Add or Subtract.

$(5 + 6i) + (6 - 3i)$

$(3.5 + 7i) - (1 + 4i)$

415

Which properties apply to complex numbers as well as real numbers?

EXAMPLE 7 Multiply.

$5i(3 + 2i)$

A complex number can be a _____.

To multiply two binomials, use _____.

EXAMPLES 8 & 9 Multiply.

$(7 - 6i)(2 + 3i)$

$(5 + 3i)(5 - 3i)$

The expressions $(a + b)$ and $(a - b)$ are called _____. The product of conjugates is always a _____.

Remember, $(a + b)(a - b) = $ _____.

Dividing complex numbers is exactly like _____ the denominator.

EXAMPLE 10 Divide

$$\frac{7 + i}{3 - 2i}$$

CONCEPT CHECK:

When $(4 - 3i)(1 + 2i)$ is multiplied by FOIL, what will the "L" term simplify to?

GUIDED EXAMPLES:

1. Simplify. $\sqrt{-36}$

2. Evaluate the powers of i. i^{13}

3. Add. $(11 + 2.4i) + (6 + 1.6i)$

4. Multiply. $(3i + 2)(4i + 1)$

Name: _____   Date: _____

Instructor: _____   Section: _____

The Discriminant in the Quadratic Formula

For all quadratic equations of the form $ax^2 + bx + c = 0$, state the quadratic formula.

_____ can be expressed as a _____ of two integers. When written in decimal form, rational numbers are _____ or _____ decimals.
Examples.

_____ are non-terminating, non-repeating decimals.
Examples.

The _____ i is defined as _____ and _____. The set of imaginary numbers consists of numbers of the form _____, where b is a real number and $b \neq 0$.

_____ are numbers of the form _____, where a is the _____ and bi is the _____.

Observe the following examples and take note of the different types of solutions.

$x^2 - 11x + 28 = 0$ Solutions: Type:

$x^2 + 5x - 12 = 0$ Solutions: Type:

$x^2 - 6x + 9 = 0$ Solutions: Type:

$x^2 + x + 5 = 0$ Solutions: Type

What is the discriminant of a quadratic equation?
What does a discriminant tell us?

If $b^2 - 4ac$ is …	Then there are …

EXAMPLES 1 & 2 Use the discriminant to determine how many and what kind
of solutions the quadratic equations will have. Do not solve the equations.
$5x^2 - 8 = 4$

$9x^2 + 24x + 16 = 0$

OTHER NOTES

EXAMPLES 3 & 4 Use the discriminant to determine how many and what kind of solutions the quadratic equations will have. Do not solve the equations.

$4x^2 + 81 = 0$

$3x^2 + 4x - 2 = 0$

CONCEPT CHECK:

What piece of the quadratic formula will result in the discriminant?

GUIDED EXAMPLES:

1. Use the discriminant to determine the number of and types of solutions to the quadratic equation. Do not solve the equation.

    $6x^2 - 3x + 4 = 0$

2. Use the discriminant to determine the number of and types of solutions to the quadratic equation. Do not solve the equation.

    $2x^2 + 10x + 8 = 0$

3. Use the discriminant to determine the number of and types of solutions to the quadratic equation. Do not solve the equation.

    $\dfrac{1}{2}x^2 - 2x - \dfrac{3}{4} = 0$

4. Use the discriminant to determine the number of and types of solutions to the quadratic equation. Then solve the equation.

    $3x^2 - 5 = 7x$

Name: _____   Date: _____

Instructor: _____   Section: _____

Notebook 33.3
Solving Quadratic Equations with Real or Complex Number Solutions

State the Quadratic Formula. Remember, this is for all quadratic equations of the form $ax^2 + bx + c = 0$, $x =$

The _____ of a quadratic equation is the simplified value of _____.

The discriminant tells the nature of the solutions to a _____ equation.

If $b^2 - 4ac$ is …	Then the solutions will be …
Positive and a perfect square	
Positive, but not a perfect square	
Zero	
negative	

The _____, i, is defined as $i = \sqrt{-1}$ and $i^2 = -1$.

Complex numbers are written in the form _____.

EXAMPLE 1 Solve by using the quadratic formula. $x^2 - 8x + 10 = 0$
Identify a, b, and c.
Substitute.
Simplify.

EXAMPLE 2 Solve by using the quadratic formula. $x^2 - 5x - 24 = 0$
Identify a, b, and c.
Substitute.
Simplify.

EXAMPLE 3 Solve by using the quadratic formula. $x^2 - 2x + 8 = 0$
Identify a, b, and c.
Substitute.
Simplify.

EXAMPLE 4 Solve by using the quadratic formula. $3x^2 + 6 = 4x$
Write in standard form.
Identify a, b, and c.
Substitute.
Simplify.

EXAMPLE 5 Solve by using the quadratic formula. $8x^2 - 4x + 1 = 0$
Identify a, b, and c.
Substitute.
Simplify.

CONCEPT CHECK:
Identify a, b, and c for the equation $3x^2 = -2x + 1$.

GUIDED EXAMPLES:

1. Solve by using the quadratic formula. $5x^2 - 12x + 7 = 0$

2. Solve by using the quadratic formula. $3x^2 + 8x + 1 = 0$

3. Solve by using the quadratic formula. $5x^2 + 9 = 6x$

4. Solve by using the quadratic formula. $2x(x + 2) + 11 = 6$

Name: _____   Date: _____

Instructor: _____   Section: _____

Notebook 33.4
Solving Equations Quadratic in Form

Write the steps of solving a quadratic equation by factoring.
1. Make sure the equation is in _____.
2. _____, if possible.
3. Set each factor _____.
4. _____ the resulting equations.
5. _____ each solution in the original equation.

Note. What happens if the quadratic expression does not factor?

What are the other options?

State the Quadratic Formula.

What does it mean if an equation is quadratic in form?

In most of these cases, what will we substitute for u?

EXAMPLE 1 Find a substitution that will make the equation quadratic, and write the resulting quadratic equation. $5x^4 + 3x^2 + 7 = 0$
Substitute.
Resulting equation.

EXAMPLES 2 & 3 Find a substitution that will make the equation quadratic, and write the resulting quadratic equation.
$2(x + 4)^2 + 6(x + 4) + 1 = 0$

$n^{-1/3} + 2n^{-1/6} - 3 = 0$

421

EXAMPLE 4 Solve. $x^4 - 10x^2 + 25 = 0$

EXAMPLE 5 Solve. $6(x-1)^2 + (x-1) - 2 = 0$

EXAMPLE 6 Solve. $x - 5x^{1/2} - 36 = 0$

CONCEPT CHECK:

For the equation $x^4 - 50x^2 + 615$, let $u = x^2$. After the substitution step, Wanda finds that $u = 25$. How would the solution be found?

GUIDED EXAMPLES:

1. Solve. $x^4 - 12x^2 + 36 = 0$

2. Solve by factoring. $x^6 + 12x^3 = 45$

3. Solve by factoring. $8(x+7)^2 + 2(x+7) - 3 = 0$

4. Solve by factoring. $x - 6x^{1/2} - 16 = 0$

Name: _____   Date: _____

Instructor: _____   Section: _____

Complex and Quadratic Applications

OTHER NOTES

A _____ in standard form $ax^2 + bx + c = 0$ can be solved in a number of ways.

To solve by factoring,
1. _____ completely.
2. Set each _____ equal to 0.
3. _____ the resulting equations to find _____.

To solve with Square Root Property,
If $x^2 = a$, then

To solve with the quadratic formula, use _____.

EXAMPLE 1 The surface area of a sphere is given by the formula $A = 4\pi r^2$. Solve for r. You do not need to rationalize the denominator.

EXAMPLE 2 Solve for n. $A = P(1 + n)^2$

What is the formula for the area of a triangle?

EXAMPLE 3 A sail with the shape of a right triangle has an area of 39 square meters. If the boom, which forms the base of the sail, is 7 meters shorter than the mast, which forms the height, find the dimensions of the sail.

EXAMPLE 4 The resistance R in an electrical circuit is given by the formula $R = \dfrac{V}{I}$, where V is the voltage, measured in volts, and I is the current in amperes. Find the resistance in a circuit where the voltage is $3 + 2i$ volts and the current is $3i$ amperes.

CONCEPT CHECK:
What is the first step to solving $T = S(3 + n)^2 - V$ for n?

GUIDED EXAMPLES:

1. Solve for x using the quadratic formula. $\quad 3khx^2 - 5x + m = 0$

2. A rectangular swimming pool has an area of 105 square yards. The length of the pool is 6 yards less than three times the width. Find the length and width. Remember that the area of a rectangle is $A = lw$.

3. The surface area of a right circular cylinder can be found with the formula $S = 2\pi rh + 2\pi r^2$. Solve for r. You do not have to rationalize the denominator.

4. Solve for y. $\quad 4(y^2 + w) - 5 = 7R$

424

Notebook 34.1
Introduction to Graphing Quadratic Functions

A _____ is an equation that may be written in the form $ax^2 + bx + c = 0$, where $a \neq 0$.

Sketch the graph of $y = x^2$.

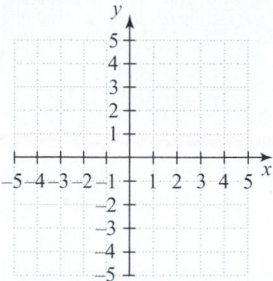

What is the vertex of a parabola?

What is the axis of symmetry of a parabola?

How do you determine if the parabola opens upward or downward?

How does that affect the vertex?

EXAMPLES 1–3 Determine whether the graph of each quadratic equation opens upward or downward.
$y = 2x^2 - 6x + 3$

$y = -5x^2 + 11$

$y = -x^2 + 4x - 1$

EXAMPLES 4–6 If an item is dropped from a height of 400 feet, its height above the ground t seconds after being dropped is given by the equation $h = -16t^2 + 400$. Determine the height of the object for the following values of t.
0 seconds

1 second

5 seconds

OTHER NOTES

CONCEPT CHECK:
What is the vertex of a parabola?

GUIDED EXAMPLES:
1. Determine whether the graph of each quadratic equation opens upward or downward.

$$y = -2x^2 + x + 5$$

2. Determine whether the graph of each quadratic equation opens upward or downward.

$$y = x^2 - 1$$

3. If a football player kicks a field goal with an initial vertical velocity of 50 feet/second, its height above the ground t seconds after being kicked is given by the equation $s = -16t^2 + 50t$. Determine the height of the football for the following value of t.   2 seconds

4. If a football player kicks a field goal with an initial vertical velocity of 50 feet/second, its height above the ground t seconds after being kicked is given by the equation $s = -16t^2 + 50t$. Determine the height of the football for the following value of t.   3 seconds

Notebook 34.2
Finding the Vertex of a Quadratic Function

A _____ is an equation that may be written in the form $ax^2 + bx + c = 0$, where $a \neq 0$.

What is the vertex of a parabola?

When will the parabola open upward?
Where is the vertex located?

When will the parabola open downward?
Where is the vertex located?

Sketch the graphs of each of these situations.

$$a > 0 \qquad\qquad a < 0$$

Write the steps for finding the vertex of a quadratic equation.
1.

2.

3.

Note: Which variable determines if the parabola opens upward or downward? Explain this.

EXAMPLE 1 Find the vertex of the quadratic equation. Is the vertex a maximum or a minimum? $y = x^2 - 8x + 15$

Identify a, b, and c.
Find x.
Substitute to find y.
Determine if vertex is a maximum or a minimum.

EXAMPLE 2 Find the vertex of the quadratic equation. Is the vertex a maximum or a minimum? $y = 12x - 3x^2 - 6$

427

EXAMPLE 3 Find the vertex of the quadratic equation. Is the vertex a maximum or a minimum? $y = 5x^2 - 7$

CONCEPT CHECK:

What is the formula for finding the x-coordinate of the vertex for $y = 2x^2 + 20x - 1$?

GUIDED EXAMPLES:

1. Find the vertex of the quadratic equation. Is the vertex a maximum or a minimum?

   $y = x^2 - 6x + 5$

2. Find the vertex of the quadratic equation. Is the vertex a maximum or a minimum?

   $y = 3x^2 + 12x + 7$

3. Find the vertex of the quadratic equation. Is the vertex a maximum or a minimum?

   $y = -4x^2 - 2x + 1$

4. Find the vertex of the quadratic equation. Is the vertex a maximum or a minimum?

   $y = -2x^2 + 10$

428

Notebook 34.3
Finding the Intercepts of a Quadratic Function

The point on the graph where a line or curve crosses an axis is called an

_____.

What is the point on the graph where the curve crosses the x-axis?
What is the ordered pair that represents that point?

How do you find any x-intercepts?

What is the point on the graph where the curve crosses the y-axis?
What ordered pair represents that point?

How do you find any y-intercepts?

How many x-intercepts will a quadratic function have?

How many y-intercepts will a quadratic function have?

Sketch the graphs that demonstrate this. Explain.

How do you find the y-intercept of a quadratic function?

EXAMPLES 1 & 2 Find the y-intercept of each quadratic function.
$f(x) = x^2 + 6x + 15$

$f(x) = 3x^2$

How do you find the x-intercept(s) of a quadratic function?

EXAMPLES 3 & 4 Find the x-intercept(s) of each quadratic function, if any exist.

$f(x) = x^2 + 2x - 24$

$f(x) = x^2 + 1$

EXAMPLE 5 Find the x- and y-intercepts of the quadratic function.

$f(x) = x^2 + 5x - 14$

Find the x-intercepts.

Find the y-intercepts.

CONCEPT CHECK:

How is the y-intercept of a quadratic function found?

GUIDED EXAMPLES:

1. Find the y-intercept of the quadratic function. $f(x) = x^2 + 3x - 10$

2. Find the y-intercept of the quadratic function. $f(x) = -2x^2 + 4x - 5$

3. Find the x-intercept(s) of the quadratic function, if any exist.
   $f(x) = x^2 + 17$

4. Find the x-intercept(s) of the quadratic function, if any exist.
   $f(x) = 5x^2 + 3x$

430

Notebook 34.4
Graphing Quadratic Functions Summary

The graph of a quadratic equation $y = ax^2 + bx + c$ is a _____.

The highest or lowest point on a parabola is the _____.
If $a > 0$, then …

If $a < 0$, then …

Where is the vertex of a parabola located?

The point where the curve crosses the x-axis is the _____.
How do you find the x-intercept(s)?

The point where the curve crosses the y-axis is the _____.
How do you find the y-intercept?

What is the axis of symmetry?

What does it mean to be symmetric?

What is the equation for an axis of symmetry?

EXAMPLE 1 Graph the equation.
$y = x^2$

x	y

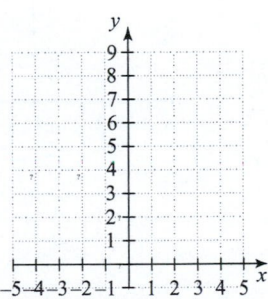

Write the steps for graphing a quadratic equation.
1. Determine the _____ of the parabola.
2. Find the _____ of the equation.
3. Find the _____ of the equation.
4. If no x-intercepts, find _____.
5. Plot the _____, _____ and _____.
6. Draw a _____ through the points.

431

EXAMPLE 2 Graph the equation.

$y = x^2 - 6x + 8$

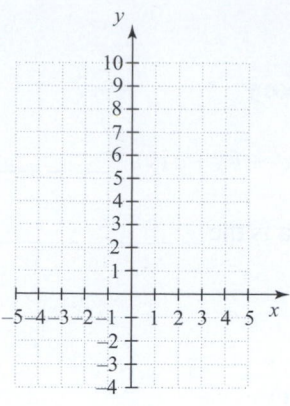

EXAMPLE 3 Graph the equation.

$y = -2x^2 + 4x - 3$

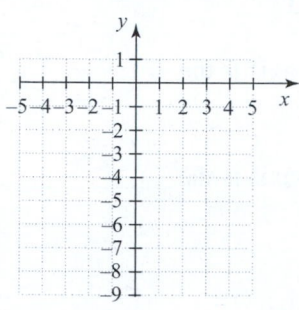

CONCEPT CHECK:

What does the equation of the axis of symmetry look like for a parabola of the form $y = ax^2 + bx + c$?

GUIDED EXAMPLES: Graph the equations.

1. $y = x^2 - 6x + 5$

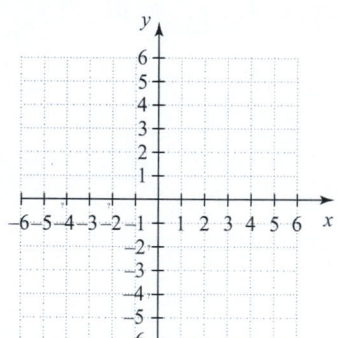

2. $y = -2x^2 - 8x - 6$

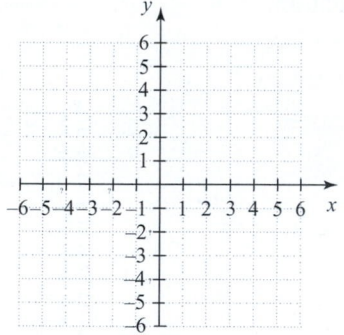

3. $y = -x^2 + 2x - 2$

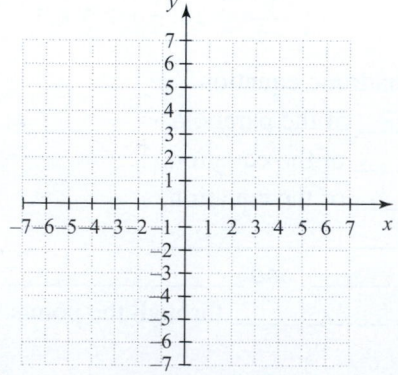

432

Name: _____   Date: _____

Instructor: _____   Section: _____

Notebook 34.5
Applications with Quadratic Functions

OTHER NOTES

The graph of a quadratic equation $y = ax^2 + bx + c$ is a _____.

The highest or lowest point on a parabola is the _____.
If $a > 0$, then …

If $a < 0$, then …

Where is the vertex of a parabola located?

What is optimization?

Write the procedure for finding the maximum or minimum value of a parabola.

Then sketch the possibilities.

EXAMPLE 1 The cost C in dollars of producing x bicycles at a factory can be modeled by the function $C(x) = 2x^2 - 760x + 85,000$. Find the number of bicycles that must be manufactured to minimize the cost. Then minimize the cost. Then find the minimum cost.

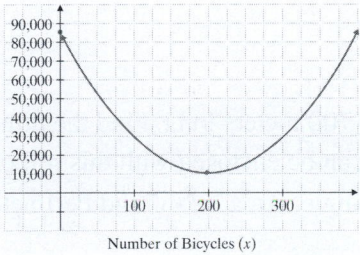

Number of Bicycles (x)

EXAMPLE 2 A roll of chain-link fence is 500 feet long. What are the maximum dimensions of a rectangular field that can be enclosed with this fence? What is the maximum are of this field?

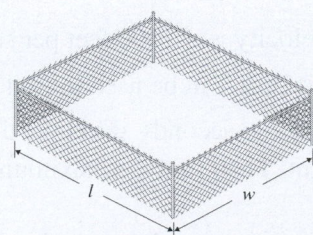

433

EXAMPLE 3 A rocket is fired upward with an initial velocity v of 110 meters per second. The quadratic function $h(t) = .05t^2 + 110t$ can be used to find the height $h(t)$ of the rocket, in meters, at any time t in seconds. Find the maximum height of the rocket and when it occurs.

CONCEPT CHECK:

What does it mean to optimize a quadratic function?

GUIDED EXAMPLES:

1. The cost C in dollars of producing n DVD players at a factory can be modeled by the function $C(n) = 3n^2 - 852n + 63,000$. Find the number of DVD players that must be manufactured to minimize the cost. Then find the minimum cost.

2. A roll of barbed-wire fencing is 700 meters long. Find the dimensions of the largest rectangular pasture that can be enclosed with this roll of barbed wire. Remember that for a rectangle, Area $= l \cdot w$ and Perimeter $= 2l + 2w$.

3. A rocket is fired upward with a velocity v of 1200 feet per second. The quadratic equation $h(t) = -16t^2 + 1200t$ can be used to find the height $h(t)$ of the rocket, in meters, at any time t in seconds. Find the time it takes the rocket to reach its maximum height. Then find the maximum height.

Notebook 35.1
Interval Notation

What symbol represents each of the following phrases?
"is less than"
"is than or equal to"
"is greater than"
"is greater than or equal to"

What does an open circle show on a graph?
What does a closed circle show on a graph?

What does the graph of an inequality represent?
How do you graph linear inequalities?

Example.
$x > 3$

_____ is another way to represent the solution to an
inequality.
What two values does interval notation use?

Note: When do you use a parentheses in interval notation and when do you use
a bracket?

Example.

Caution! Interval notation is not the same as an _____!

An _____ is a set whose elements cannot be counted.

What is infinity?

What is negative infinity?

When an interval contains all the points to ∞ or $-\infty$, you must use it as one
of the _____. These symbols get _____ in interval
notation.

Example. $x > 3$

435

EXAMPLES 1 & 2 Graph each inequality and write in interval notation.

$x < -2$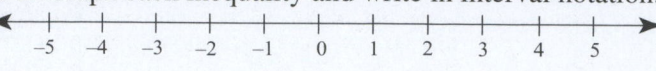

$x \geq 0$

Inequality	Graph	Interval Notation
$x > a$		
$x \geq a$		
$x < a$		
$x \leq a$		

Note: A parentheses is always used to enclose _____ and _____.

Inequality	Graph	Interval Notation
$a < x < b$		
$a \leq x \leq b$		
$a \leq x < b$		
$a < x \leq b$		

EXAMPLES 3–6 Graph each inequality and write in interval notation.

$-1 \leq x < 4$ $-3 \leq x \leq 0$

$-2 \leq x < 2$ $-1.5 < x \leq 3$

CONCEPT CHECK:

Describe the graph of $(-\infty, 3]$.

GUIDED EXAMPLES: Graph each inequality and write the interval notation.

1. $n \leq 0$

2. $x < -1.5$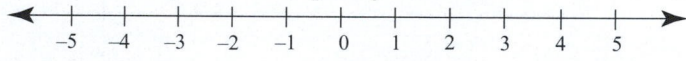

3. $1 < x \leq 3.5$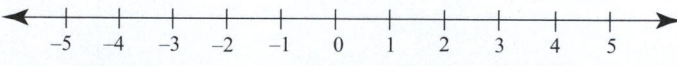

4. $-8 \leq x < -4.1$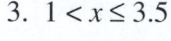

436

Name: _____   Date: _____

Instructor: _____   Section: _____

Notebook 35.2
Graphing Compound Inequalities

What is the difference between an "and" statement and an "or" statement?

$x > -2$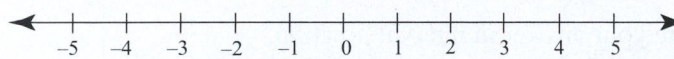

$x \leq 4$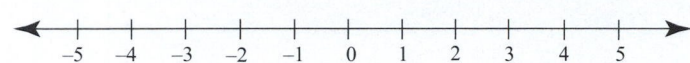

EXAMPLE 1 Graph. Write your answer in interval notation.

$x > -2$ and $x \leq 4$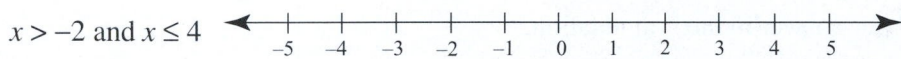

What is a compound inequality?

Write the steps for graphing a compound inequality.
1. _____ the first inequality.
    a.
    b.
    c.
2. _____ the second inequality.
3. Graph the _____ to the _____.

Note: Where do you shade for "and" and where do you shade for "or"?

What are some different ways to write "and" inequalities?

EXAMPLES 2–4 Graph.

$-3 \leq x < 3$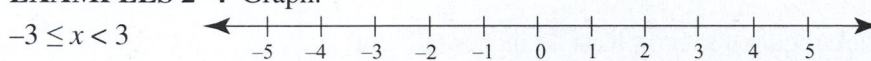

$(-2, 5]$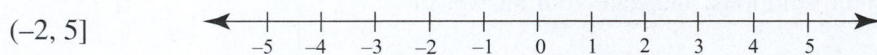

$x > 4$ and $x < 1$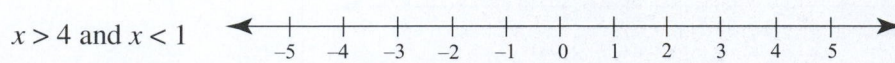

437

OTHER NOTES

EXAMPLE 5 The U.S. Air Force requires female pilots to be at least 58 inches tall, but they cannot be more than 80 inches tall. Write a compound inequality in set notation and in interval notation. Then graph the solutions.

EXAMPLES 6 & 7 Graph. Write your answer in interval notation.

$x > 3$ or $x \leq -2$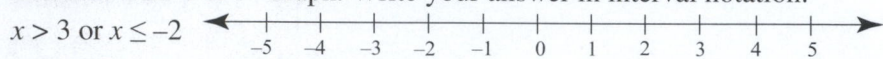

$x \geq -1$ or $x \leq 5$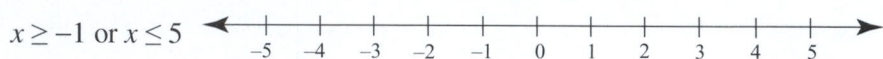

EXAMPLE 8 Graph. Write your answer in interval notation.

$n < 4$ and $n > -2$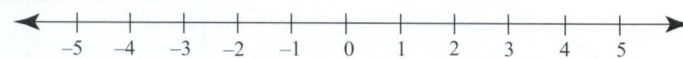

EXAMPLE 9 Graph. Write your answer in interval notation.

$x < 5$ or $x > -3$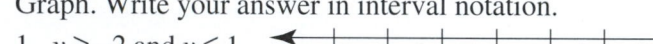

CONCEPT CHECK:

A talent show states that the contestants must be at least 16 years of age, but no greater than 28 years of age. How will this be written as an inequality?

GUIDED EXAMPLES:

Graph. Write your answer in interval notation.

1. $y \geq -2$ and $y \leq 1$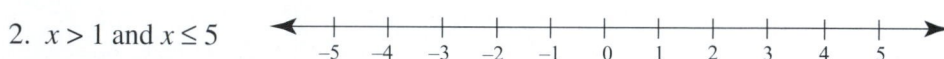

2. $x > 1$ and $x \leq 5$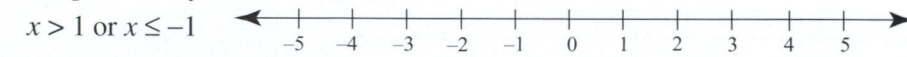

3. To ride a roller coaster, riders are required to be at least 44 inches tall, but they cannot be more than 81 inches tall. Write a compound inequality to represent this situation. Graph the solutions, and state your answer in interval notation.

4. Graph. Write your answer in interval notation.

   $x > 1$ or $x \leq -1$

438

Name: _____   Date: _____

Instructor: _____   Section: _____

Notebook 35.3
Solving Compound Inequalities

Graph.

$x > -2$ and $x \leq 3$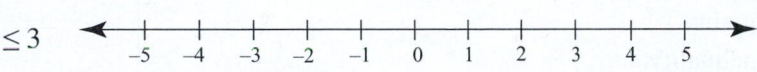

EXAMPLE 1 Solve and graph. Then write your answer in interval notation.

$x + 1 > -2$ and $x - 2 \leq 2$

EXAMPLE 2 Solve and graph. Then write your answer in interval notation.

$3x + 2 \geq 14$ or $2x - 1 \leq -7$

Solve and graph the first inequality.
Solve and graph the second inequality.
Graph the solution to the compound inequality.
Write in interval notation.

EXAMPLE 3 Solve and graph. Then write your answer in interval notation.

$3x + 5 \leq 17$ or $-4x > 12$

Solve and graph the first inequality.
Solve and graph the second inequality.
Graph the solution to the compound inequality.
Write in interval notation.

439

EXAMPLE 4 Solve and graph. Then write your answer in interval notation.

$6 + x < 9$ or $5x + 2 > -18$
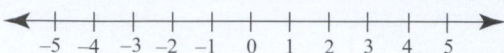

Solve and graph the first inequality.
Solve and graph the second inequality.
Graph the solution to the compound inequality.
Write in interval notation.

EXAMPLE 5 Solve and graph. Then write your answer in interval notation.

$-9 \le 3a - 3 < 9$
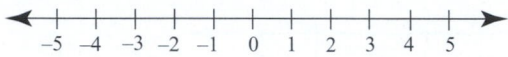

Solve and graph the first inequality.
Solve and graph the second inequality.
Graph the solution to the compound inequality.
Write in interval notation.

EXAMPLE 6 Solve and graph. Then write your answer in interval notation.

$12 < -2x + 3 < 1$
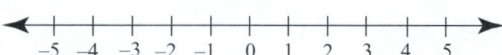

Solve the first inequality.
Solve the second inequality.

CONCEPT CHECK:
How would the solution to the compound inequality "$x + 1 > 3$ and $2x \le 8$" be written in interval notation?

GUIDED EXAMPLES: Solve and graph. Then write the solution in interval notation.

1. $2x + 3 \le 5$ and $-x - 1 \le 2$
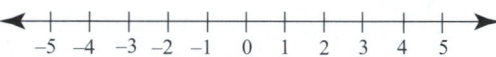

2. $2n + 3 > -7$ or $7n - 1 > 13$
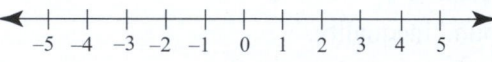

3. $3x + 9 > -6$ or $7x + 6 \le 20$
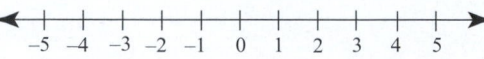

4. $-40 \le 8m - 16 \le -8$
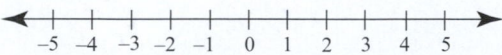

440

Name: _____   Date: _____

Instructor: _____   Section: _____

Notebook 35.4
Solving Quadratic Inequalities

Observe the graph of $f(x) = x^2 - 6x + 8$

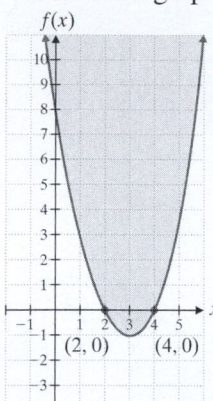

Write the steps for solving a quadratic equation.
1. Make sure the equation is in _____.
2. Solve by _____ or the _____.
3. Check your solutions.

Write the steps for solving a quadratic inequality.
1. Replace the _____ with an _____.
   Solve to find the _____ (or _____).

2. Use the _____ to separate the _____ into
   distinct _____. Mark the _____ on a number
   line.

3. Choose a _____ in each _____.

4. _____ each of these numbers in the _____.
   If the test number makes the inequality _____, _____ the
   region in which it is located; if the test number does _____ make
   the inequality _____, _____ shade that region.

5. _____ which regions satisfy the original _____ of
   the quadratic inequality.

EXAMPLE 1 Solve and graph the solution. Write the solution in interval
notation.

$(x - 3)(x + 1) > 0$

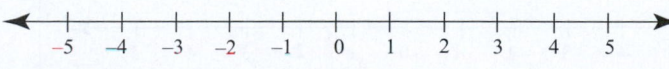

441

EXAMPLE 2 Solve and graph the solution. Write the solution in interval notation.

$2x^2 + x - 6 \leq 0$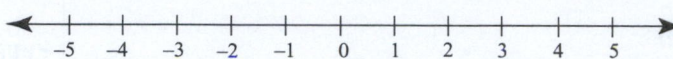

EXAMPLE 3 Solve and graph the solution. Write the solution in interval notation.

$x^2 - 4x + 4 > 0$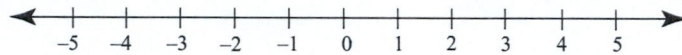

EXAMPLE 4 Solve and graph the solution. Write the solution in interval notation.

$x^2 + 4x - 6 \geq 0$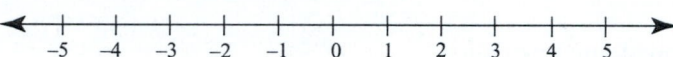

CONCEPT CHECK:
For the inequality $(x + 3)(x - 2) < 0$, what are the test regions?

GUIDED EXAMPLES: Solve and graph. Write the solution in interval notation.

1. $x^2 - 2x - 8 > 0$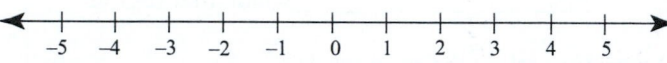

2. $4x^2 - 11x - 3 < 0$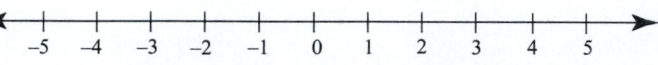

3. $x^2 + 2x - 7 \leq 0$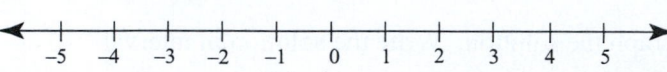

4. $x^2 - 4x > 0$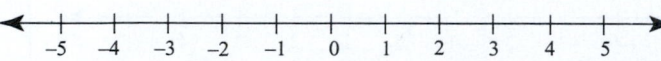

442

Name: _____   Date: _____

Instructor: _____   Section: _____

Notebook 35.5
Solving Rational Inequalities

Solving an inequality containing a rational expression is very similar to solving a _____.

Write the steps for solving rational inequalities.

1.

2.

3.

4.

5.

6.

EXAMPLE 1 Solve and graph the solution. Write the solution in interval notation.

$$\frac{(x-3)}{(x+1)} < 0$$

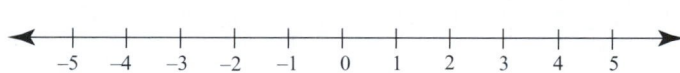

EXAMPLE 2 Solve and graph the solution. Write the solution in interval notation.

$$\frac{x^2 + 8x + 15}{x-1} \le 0$$

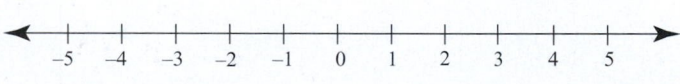

EXAMPLE 3 Solve and graph the solution. Write the solution in interval notation.

$\dfrac{x+9}{x}+2\le 0$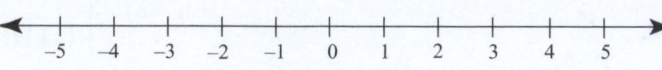

CONCEPT CHECK:

What are the boundary points for the rational inequality $\dfrac{x^2+2x-3}{x+2}>0$?

GUIDED EXAMPLES: Solve and graph. Write the solutions in interval notation.

1. $\dfrac{x+2}{x-4}>0$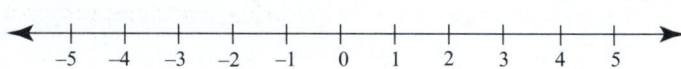

2. $\dfrac{x^2+x-6}{x+1}>0$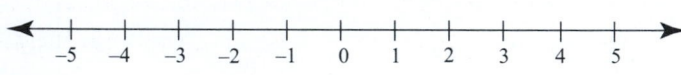

3. $\dfrac{2x-6}{x}+2\ge 0$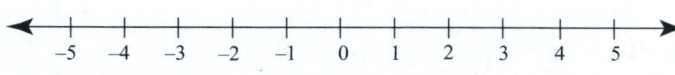

444

Name: _____   Date: _____

Instructor: _____   Section: _____

Notebook 36.1
Introduction to Absolute Value Equations

The _____ of a number is the distance between that number and zero on a number line.

EXAMPLE 1 Solve. $|x| = 2$

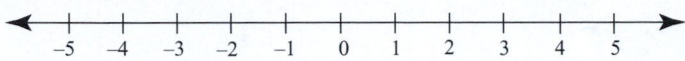

EXAMPLES 2–4 Solve.
$|x| = 8.35$

$|y| = 0$

$|x| = -11$

EXAMPLE 5 Solve. $|x| + 1 = 10$

EXAMPLE 6 Solve. $-6|x| = -72$

445

EXAMPLES 7 & 8 Write an equation using absolute value to represent the given graph.

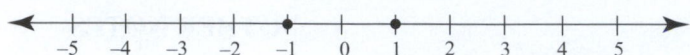

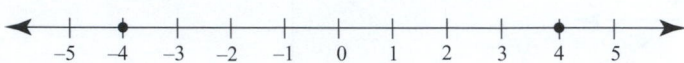

CONCEPT CHECK:
What is the FIRST step to solving $2|x| - 3 = -2$?

GUIDED EXAMPLES: See guided examples
1. Solve. $|y| = 7$

2. Isolate the absolute value, then solve. $|a| - 18 = -6$

3. Isolate the absolute value, then solve. $-3|x| = -5.4$

4. Write an equation using absolute value to represent the given graph.

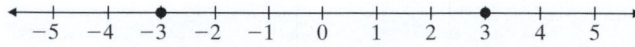

Notebook 36.2
Solving Basic Absolute Value Equations

OTHER NOTES

If $|x| = a$, and a is a positive real number, then _____.
The _____ of a number is the _____
between that number and _____ on a number line.

Example. If $|x| = 4$, then $x = 4$ or $x = -4$.

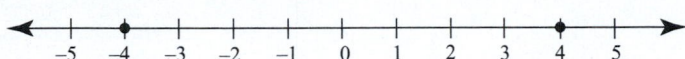

Write the steps for solving absolute value equations.
1.

2.

3.

EXAMPLE 1 Solve. $|x + 1| = 6$
Write as two separate equations.

Solve the resulting equations.

Check.

EXAMPLE 2 Solve. $|5y| = 35$

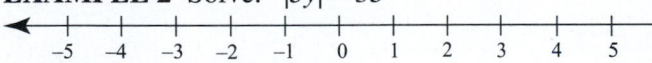

Write as two separate equations.

Solve the resulting equations.

Check.

EXAMPLE 3 Solve. $\left|\dfrac{1}{2}x - 1\right| = 5$

447

EXAMPLES 4 Solve. $|3x - 1| + 2 = 5$

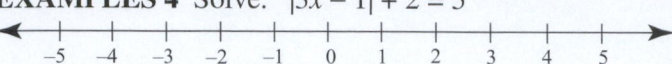

Isolate the absolute value.

Write as two equations.

Solve.

Check.

EXAMPLE 5 Solve. $|8x - 3| + 12 = 7$

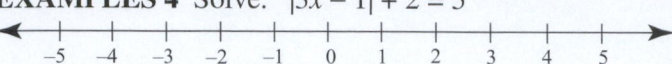

Caution! An absolution value can never equal a _____!

CONCEPT CHECK:
What are the two equations that will be solved for $|2x + 6| = 10$?

GUIDED EXAMPLES: Solve.

1. $|x + 5| = 17$

2. $|7x - 3| = 11$

3. $\left|\dfrac{1}{2} - \dfrac{3}{4}x\right| = 2$

4. $|2x + 1| - 3 = 2$

448

Notebook 36.3
Solving Multiple Absolute Value Equations

If $|x| = a$, and a is a positive real number, then _____ or _____.
What must happen in order for two absolute value expressions to be equal?

To solve multiple absolute value equations, use the rules below.
If $|x| = |a|$, then either …

If $|ax + b| = |cx + d|$, then either …

Write the steps for solving absolute value equations.
1. _____ the absolute value equation as two _____.

2. _____ the resulting equations.

3. _____.

EXAMPLE 1 Solve. $|3x + 4| = |x|$
Write as two separate equations.

Solve the resulting equations.

Check.

EXAMPLES 2 Solve. $|3x - 4| = |x + 6|$
Write as two separate equations.

Solve the resulting equations.

Check.

EXAMPLE 3 Solve. $|x + 3| = |x - 5|$

EXAMPLE 4 Solve. $|2x - 4| = |4 - 2x|$

CONCEPT CHECK:
When rewriting $|x + 2| = |2x - 5|$, one equation that will be solved is
$x + 2 = 2x - 5$. What is the other one?

GUIDED EXAMPLES: Solve.

1. $|9x - 40| = |x|$

2. $|x - 6| = |5x + 8|$

3. $|3 - x| = \left|\dfrac{x}{2} + 3\right|$

4. $|2x + 1| = |2x - 9|$

450

Notebook 36.4
Solving Absolute Value Inequalities

OTHER NOTES

Consider $|x| < 4$.

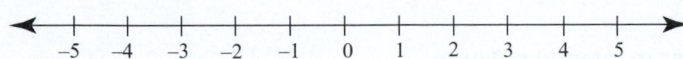

What does it mean if $|x| < a$?
In interval notation, if $|x| < a$, then …

Consider $|x| > 4$.

What does it mean if $|x| > a$?
In interval notation, if $|x| > a$, then …

EXAMPLES 1 & 2 Graph. Write the answer in interval notation.

$|x| \geq 2$

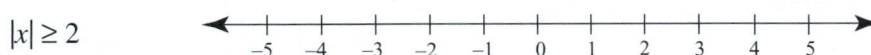

$|x| < 3$

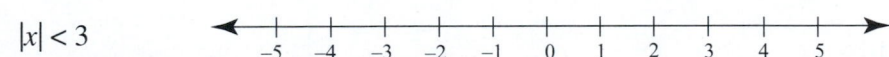

How do you know when to use parentheses or brackets?

EXAMPLE 3 Solve and graph. Write the answer in interval notation.
$|x + 5| \leq 10$
Rewrite according to appropriate rule.
Solve.
Graph and write interval notation.

451

EXAMPLE 4 Solve and graph. Write the answer in interval notation.
$|-2x - 1| \geq 7$

EXAMPLE 5 Solve and graph. Write the answer in interval notation.
$|4x + 2| + 5 > 9$

Isolate the absolute value.

Rewrite according to appropriate rule.

Solve.

Graph and write interval notation.

EXAMPLE 6 Solve and graph. Write the answer in interval notation.
$|x| > -2$

The solution set is _____.

EXAMPLE 7 Solve and graph. Write the answer in interval notation.
$|x| \leq -6$

This inequality has _____.

CONCEPT CHECK:

What is the interval notation for $|x| \geq 3$?

GUIDED EXAMPLES:

1. Graph on a number line. $|x| \geq 4$

2. Solve and graph. Write the solution in interval notation. $|x + 3| \leq 7$

3. Solve and graph. Write the solution in interval notation. $|4x - 3| \leq 9$

4. Solve and graph. $|3x + 5| + 9 \geq 26$

452

Name: _____ Date: _____

Instructor: _____ Section: _____

Notebook 37.1
Introduction to Conic Sections

What is a conic section?

Name the four types of conic sections we will be discussing.
1.
2.
3.
4.

Define circle.

What is the fixed distance called?
How is a circle formed?

What is the form of an equation of a circle with center at $(0, 0)$ and radius r?
What is the equation of a circle with center (h, k) and radius r?

Define parabola.

Sketch the graph of a parabola.

How is a parabola formed?

What is the standard form of the equation of a vertical parabola with its vertex at the origin?
What is the standard form of the equation of a vertical parabola with its vertex at (h, k)?
What is the standard form of the equation of a horizontal parabola with its vertex at the origin?
What is the standard form of the equation of a horizontal parabola with its vertex at (h, k)?

Define ellipse.

Sketch the graph of an ellipse.

What are the foci?
How is an ellipse formed?

What is the equation of an ellipse with center at the origin?
What is the equation of an ellipse with center at (h, k)?

Define hyperbola.

What are the fixed points called?
How is a hyperbola formed?

What is the equation of a vertical hyperbola with center at the origin?
What if the center is at (h, k)?
What is the equation of a horizontal hyperbola with center at the origin?
What if the center is at (h, k)?

EXAMPLES 1 & 2 Identify the conic sections shown.

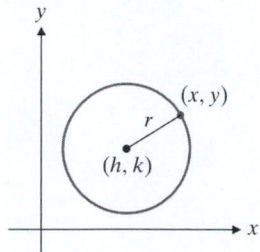 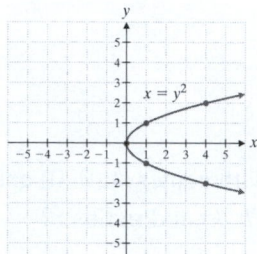

EXAMPLES 3 & 4 Identify the conic sections shown.

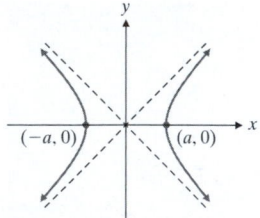 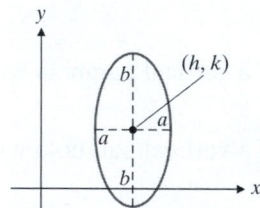

EXAMPLES 5–8 Identify the conic sections described as a circle, a parabola, an ellipse, or a hyperbola.

For each point in the set, the absolute value of the difference of its distance to two fixed points is constant.

For each point in the set, the sum of its distances to two fixed points is constant.

Each point in the set is a fixed distance from a center point.

This is the graph of a quadratic equation.

454

Name: _____   Date: _____

Instructor: _____   Section: _____

EXAMPLES 9–12 Identify each conic section from its equation as a circle, a parabola, and ellipse, or a hyperbola.

$x = -3(y - 5)^2 + 7$

$$\frac{(y-2)^2}{3^2} - \frac{(x+6)^2}{11^2} = 1$$

$$\frac{(x+3)^2}{1^2} - \frac{(y+8)^2}{9^2} = 1$$

$(x + 2)^2 + (y - 6)^2 = 5^2$

CONCEPT CHECK:

What shape does the equation $(x - h)^2 + (y - k)^2 = r^2$ represent?

GUIDED EXAMPLES: Identify each conic section as a circle, a parabola, an ellipse, or a hyperbola.

1. This is formed if a plane cuts a cone at an angle so that the plane intersects all sides of the cone, and is not parallel to the base of the cone.

2.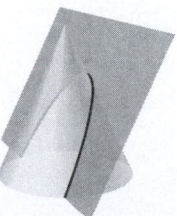

3.

4. $y = -4(x - 7)^2 + (-2)$

Name: _____   Date: _____

Instructor: _____   Section: _____

Notebook 37.2
The Circle
What is a conic section?
What are the four basic shapes of a conic section?
1.
2.
3.
4.

Define a circle.
What is the radius?

To calculate the distance between any two point (x_1, y_1) and (x_2, y_2),
what is the formula to use? $D =$
Let a circle of radius r have its center at (h, k).
For any point (x, y) on the circle, how do you find r?
Write the equation of a circle with center at (h, k) and radius r.

Note: What is the equation if the center of the circle is at $(0, 0)$?

EXAMPLE 1 Find the center and radius of the following circle. Then sketch
its graph.

$x^2 + y^2 = 9$

Find the center.
Find the radius.
Graph.

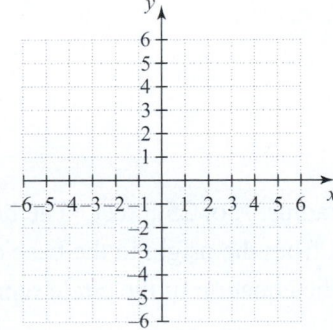

EXAMPLE 2 Find the center and radius of the following circle. Then sketch
its graph.

$(x+2)^2 + (y-3)^2 = 25$

Rewrite in standard form.

Find the center.
Find the radius.
Graph.

Note: If the equation is not written in standard form, what do you do?

457

EXAMPLE 3 Find the center and radius of the following circle. Then sketch its graph.

$$x^2 + 2x + y^2 + 6y + 6 = 0$$

Write in standard form.

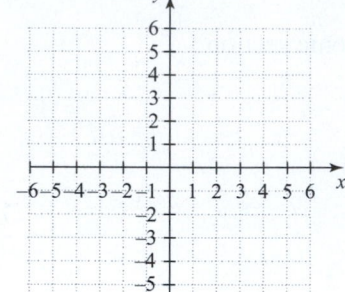

Find the center.
Find the radius.
Graph.

EXAMPLES 4–6 Write the equation of the circles in standard form with the given centers and radii.

Center $(0, 0)$; $r = 2$

Center $(8, -2)$; $r = 7$

Center $(-1, 3)$; $r = \sqrt{5}$

EXAMPLE 7 A Ferris wheel has a radius, r, of 25.3 feet. The height of the tower, t, is 31.8 feet. The distance, d, from the origin to the base of the tower is 44.8 feet. Find the standard form of the equation of the circle represented by the Ferris wheel.

CONCEPT CHECK:
What is a conic section?

458

Name: _____   Date: _____

Instructor: _____   Section: _____

GUIDED EXAMPLES:

1. Find the center and radius of the following circle. Then sketch its graph.

   $x^2 + y^2 = 25$

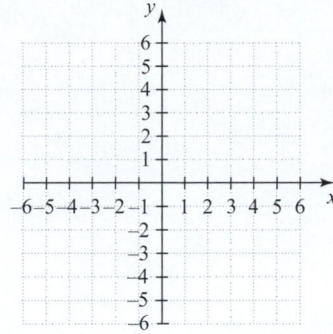

2. Find the center and radius of the following circle. Then sketch its graph.

   $(x-3)^2 + (y-2)^2 = 4$

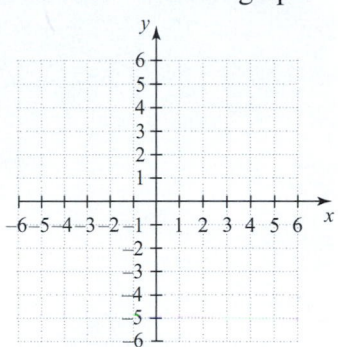

3. Write the equation of the circle in standard form. Find the radius and center of the circle. Then sketch its graph.

   $x^2 + 4x + y^2 + 2y - 20 = 0$

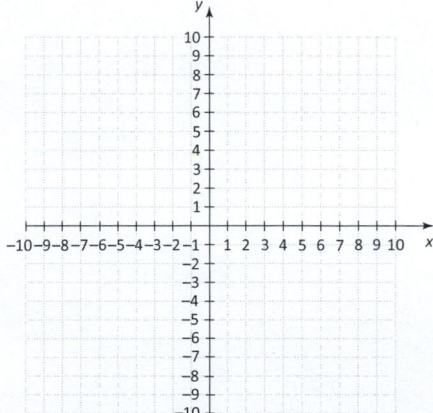

4. Write the equation of the circle in standard form with the given center and radius.   Center $(-3, 7)$; $r = 6$

459

Name: _____   Date: _____

Instructor: _____   Section: _____

Notebook 37.3
The Parabola

A _____ is a shape that can be formed by slicing a cone with a plane.

A _____ is the graph of a quadratic equation.
What is the general shape of a parabola?

What is the simplest form for the equation of a parabola?

What is the difference between these two types?

Note: What does it mean if the graph of $y = x^2$ is symmetric about the y-axis?

The _____ is the line over which the parabola is symmetric.

The _____ is the point at which the parabola crosses the axis of symmetry. This is either the _____ or _____ point on the parabola.

Write the standard form of the equation of a vertical parabola.
1. The graph of _____ is a vertical parabola.
2. The parabola opens _____ if a ___ 0 and downward if a ___ 0.
3. The _____ of the parabola is _____.
4. The _____ is the line, $x =$ _____.
5. The _____ is the point where the parabola crosses the _____, that is, where _____.

Note: If we know a point on the parabola, how can we find another point?

EXAMPLE 1 Graph the parabola. $y = 2(x - 1)^2 - 3$

Determine if vertical or horizontal.
Identify a, h, and k.
Determine the direction.
Find the vertex.
Find axis of symmetry.
Find the y-intercept.
Graph.

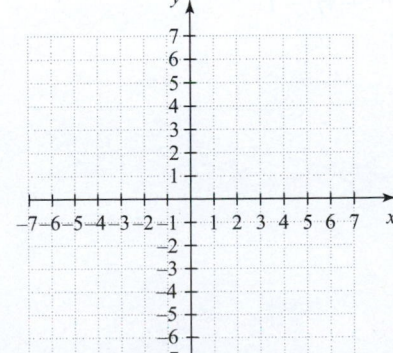

461

Write the standard form of the equation of a horizontal parabola.

1. The graph of _____ is a horizontal parabola.

2. The parabola opens to the _____ if a ___ 0 and to the
   _____ if a ___ 0.

3. The _____ of the parabola is _____.

4. The _____ is the line, $y =$ _____.

5. The _____ is the point where the parabola crosses the
   _____, that is, where _____.

Note: If we know a point on the parabola, how can we find another point?

EXAMPLE 2 Graph the parabola. $x = (y + 2)^2 - 3$

Determine if vertical or horizontal.

Identify a, h, and k.

Determine the direction.

Find the vertex.

Find axis of symmetry.

Find the x-intercept.

Graph.

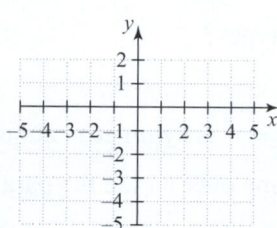

CONCEPT CHECK:

True/False   The axis of symmetry is part of the graph of a parabola.

GUIDED EXAMPLES:

1. Graph the parabola.

   $y = (x - 2)^2 + 3$

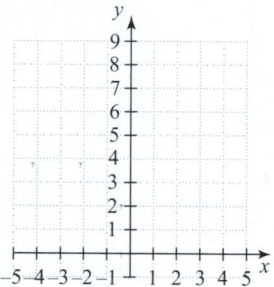

2. Graph the parabola.

   $x = -(y + 2)^2 + 1$

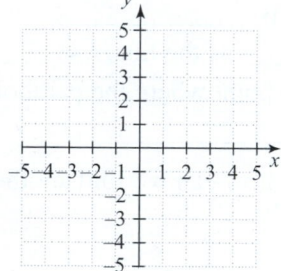

3. Graph the parabola. $y = -\dfrac{1}{2}(x + 3)^2 - 1$

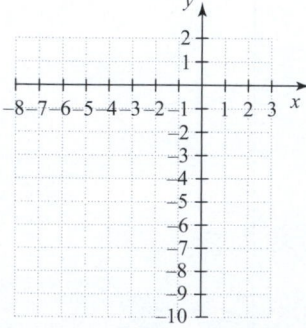

Notebook 37.4
The Ellipse

A _____ is a shape that can be formed by slicing a cone with a plane.

What is an ellipse?

What are the foci?

What is the standard form of the equation of an ellipse with center at the origin?

What are the vertices of this ellipse?

EXAMPLE 1 Graph the ellipse. Label the intercepts.

$$\frac{x^2}{36} + \frac{y^2}{4} = 1$$

Find the intercepts.

Plot the intercepts.

Draw the ellipse.

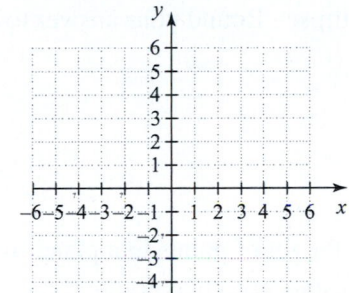

Sometimes it is necessary to rewrite the equation in

_____.

EXAMPLE 2 Graph the ellipse. Label the intercepts.

$$x^2 + 3y^2 = 12$$

Rewrite in standard form.

Find the intercepts.

Plot the intercepts.

Draw the ellipse.

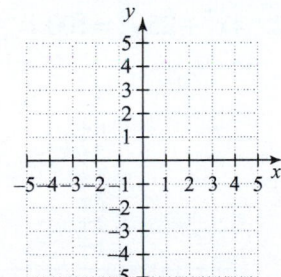

Write the standard form of the equation of an ellipse.
An ellipse with center _____ has the equation _____.

Note: The _____ is a special case of an _____.

*What happens if the center of an ellipse is not the origin?

463

EXAMPLE 3 Graph the ellipse.

$$\frac{(x-5)^2}{4}+\frac{(y-6)^2}{9}=1$$

Identify the center.

*Use *a* to plot points on the ellipse.

*Use *b* to plot points on the ellipse.

Draw the ellipse.

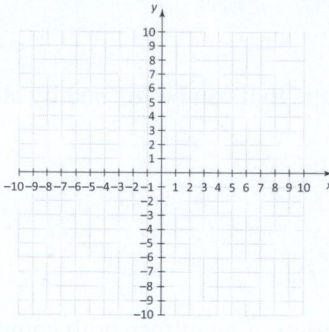

EX AMPLE 4 The orbit of Venus is an ellipse with the Sun as a focus. If we say that the center of the ellipse is at the origin, an approximate equation for the orbit is $\frac{x^2}{36}+\frac{y^2}{4}=1,$ where *x* and *y* are measured in millions of miles. Find the largest possible distance across the ellipse. Round your answer to the nearest million miles.

CONCEPT CHECK: An ellipse is the set of points in a plane such that for each point in the set, the sum of the distances to two points is constant. What are the fixed points called?

GUIDED EXAMPLES:
Graph the ellipse. Label the intercepts.

1. $\frac{x^2}{4}+\frac{y^2}{16}=1$

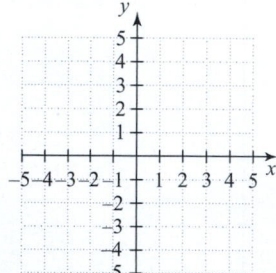

2. $4x^2+25y^2=100$

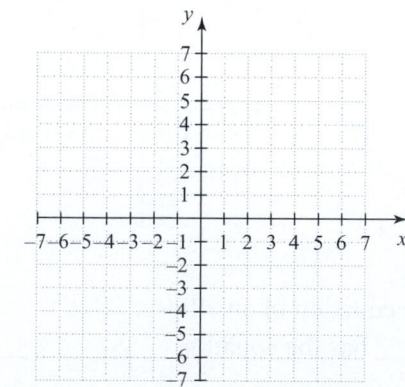

464

Name: _____   Date: _____

Instructor: _____   Section: _____

3. Graph the ellipse. $\dfrac{(x-2)^2}{16}+\dfrac{(y+3)^2}{9}=1$

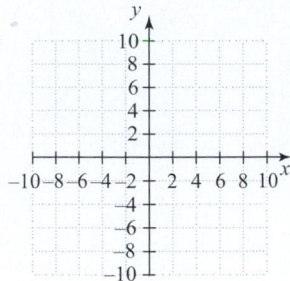

4. Bob's backyard is a rectangle, 40 meters by 60 meters. He uses the backyard for an exercise area for his dog. He drove two posts into the ground and fastened a rope to each post, passing the rope through the metal ring on his dog's collar. When the dog pulls on the rope while running, its path is an ellipse. If the dog can just reach all four sides of the rectangle, find the equation of the elliptical path.

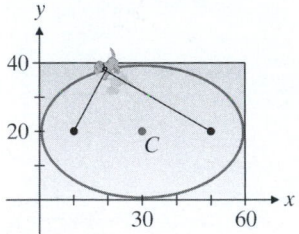

465

Name: _____   Date: _____

Instructor: _____   Section: _____

Notebook 37.5
The Hyperbola

A _____ is a shape that can be formed by slicing a cone with a plane.

What is a hyperbola?

Caution! What is the difference between a hyperbola and an ellipse?

The standard form of the equation of a hyperbola with center at the origin.
Let a and b be any positive real numbers.

A _____ with center at the _____ and vertices

_____ and _____ has the equation _____.

This is called a _____.

The axis of a horizontal hyperbola is the _____.

Sketch the graph of a horizontal hyperbola to the right.

A _____ with center at the _____ and vertices

_____ and _____ has the equation _____.

This is called a _____.

The axis of a vertical hyperbola is the _____.

Sketch the graph of a vertical hyperbola to the right.

What are the asymptotes of a hyperbola?

What are the asymptotes of the hyperbolas $\dfrac{x^2}{a^2} - \dfrac{y^2}{b^2} = 1$ and $\dfrac{y^2}{b^2} - \dfrac{x^2}{a^2} = 1$?

True/False Asymptotes are part of the graph.

The fundamental rectangle of a hyperbola is the rectangle whose center is at the _____. Where are the corners?

EXAMPLE 1 Graph the hyperbola. $\dfrac{x^2}{25} - \dfrac{y^2}{16} = 1$

Determine if vertical or horizontal.
Identify the vertices.
Draw the fundamental rectangle.
Draw the asymptotes.
Draw the hyperbola.

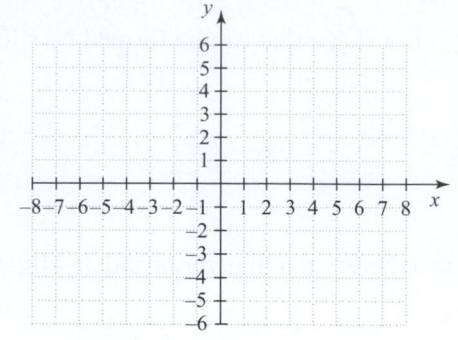

Sometimes it is necessary to rewrite the equation in _____.

EXAMPLE 2 Graph the hyperbola. $4y^2 - 7x^2 = 28$

Rewrite in standard form.
Determine if vertical or horizontal.
Identify the vertices.
Draw the fundamental rectangle.
Draw the asymptotes.
Draw the hyperbola.

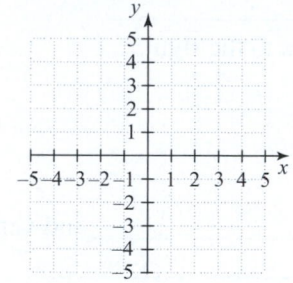

What happens if the center of an ellipse is not the origin?

Write the standard form of the equation of a horizontal hyperbola
with center at (h, k) and vertices $(h - a, k)$ and $(h + a, k)$.

Write the standard form of the equation of a vertical hyperbola
with center at (h, k) and vertices $(h, k + b)$ and $(h, k - b)$.

EXAMPLE 3 Graph the hyperbola. $\dfrac{(x-4)^2}{9} - \dfrac{(y-5)^2}{4} = 1$

Determine if vertical or
horizontal.
Identify the center.
Identify the vertices.
Draw the fundamental
rectangle.
Draw the asymptotes.
Draw the hyperbola.

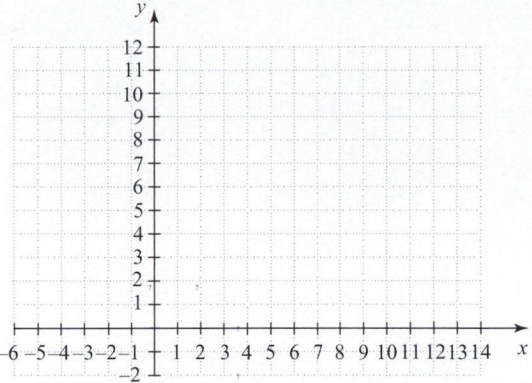

CONCEPT CHECK: The set of points in a plane such that for each point in the set, the absolute value of the difference of its distance to fixed points is a constant is called a _____.

GUIDED EXAMPLES: Graph the hyperbola.

1. $\dfrac{x^2}{16} - \dfrac{y^2}{25} = 1$

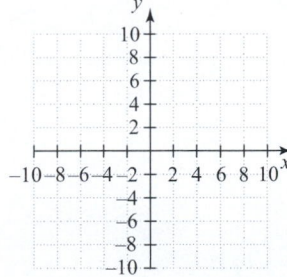

2. $3x^2 - y^2 = 9$

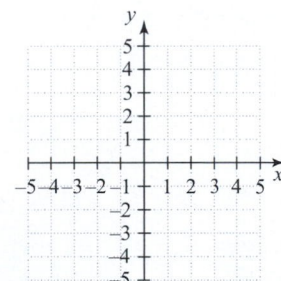

3. $\dfrac{(x+1)^2}{25} - \dfrac{(y-1)^2}{9} = 1$

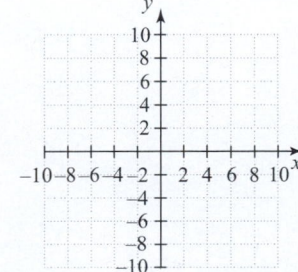

4. $\dfrac{(y+2)^2}{9} - \dfrac{(x-3)^2}{16} = 1$

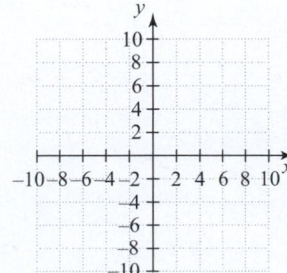

469

Name: _____ Date: _____

Instructor: _____ Section: _____

Notebook 38.1
Composite Functions

What is function composition?
How is function composition written?

How do you evaluate $(f \circ g)(x)$?

EXAMPLES 1 & 2 Given $f(x) = 2x$ and $g(x) = 4x + 7$, find the following.
$f[g(x)]$

$g[f(x)]$

Caution! What does it mean that the composition of functions is not commutative?

EXAMPLES 3 & 4 Given $f(x) = 4x^2$ and $g(x) = 3x - 5$, find the following.
$(f \circ g)(x)$

$(g \circ f)(4)$

EXAMPLE 5 Given $f(x) = \dfrac{7}{2x - 3}$ and $g(x) = x + 2$, find the following.
$f[g(x)]$

Caution! Why must you state any excluded values?

How do you find any excluded values?

OTHER NOTES

CONCEPT CHECK:

True/False Function composition is a way of using the output of one function as the input of another function.

GUIDED EXAMPLES:

1. Given $f(x) = -9x$ and $g(x) = -2x + 3$, find the following.

   $g[f(x)]$

2. Given $f(x) = 2x^2$ and $g(x) = 3x - 7$, find $(f \circ g)(x)$.

3. Given $f(x) = -6x^2 + 1$ and $g(x) = 5x + 1$, find $(g \circ f)(x)$.

4. Given $f(x) = 3x^2$ and $g(x) = 2x - 7$, find $(g \circ f)(5)$.

Name: _____   Date: _____

Instructor: _____   Section: _____

Notebook 38.2
Inverse Functions

OTHER NOTES

A _____ is a relation for which every x value in the domain
has _____ y value. That is no two
different ordered pairs have the same _____.

Note: What is the vertical line test?

What is a one-to-one function?

EXAMPLES 1 & 2 Determine whether the relations are one-to-one functions.
$F = \{(1, -3), (2, 0), (3, 3), (4, 6), (5, 9)\}$

$G = \{(-2, 5), (-1, 2), (0, 1), (1, 2), (2, 5)\}$

What does the horizontal line test say?

EXAMPLES 3 & 4 Determine whether each relation is a one-to-one function.

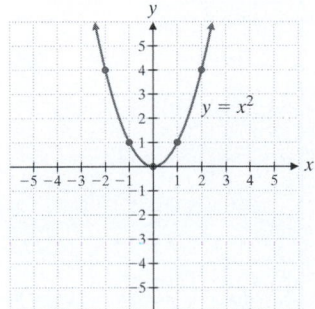
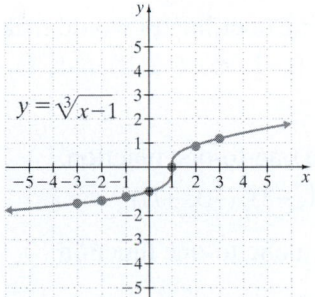

What is an inverse function?

How is an inverse function written?

EXAMPLE 5 Find the inverse of the one-to-one function.
$A = \{(1, -3), (2, 0), (3, 3), (4, 6), (5, 9)\}$

473

Write the steps for finding the inverse of a one-to-one function.

1.

2.

3.

4.

EXAMPLE 6 Find the inverse of the one-to-one function.

$f(x) = 2x + 2$

EXAMPLE 7 Find the inverse of the one-to-one function.

$f(x) = x^3 + 7$

What can be said about the graph of a function and its inverse?

EXAMPLE 8 Graph the function, then find and graph its inverse.

$f(x) = 2x + 2$

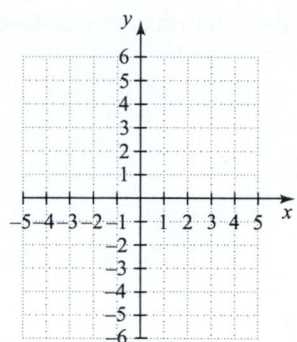

CONCEPT CHECK:

Removing which of the following coordinates would make the function G one-to-one? $G = \{(-3, -5), (-2, 0), (-1, 1), (2, 0), (1, 3)\}$

GUIDED EXAMPLES:

1. Determine if the relation is a one-to-one function.

    $S = \{(5, 9), (3, 5), (1, 1), (-1, -3), (-3, -7)\}$

2. Determine if the function is a one-to-one function.

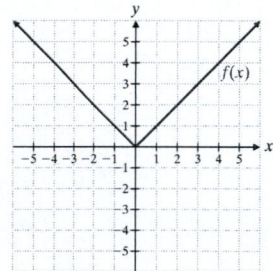

3. Find the inverse of the one-to-one function.

    $A = \{(3, -1), (8, 0), (13, 1), (18, 2), (23, 3)\}$

4. Find the inverse of the one-to-one function. $f(x) = 3x$

474

Name: _____   Date: _____

Instructor: _____   Section: _____

Evaluating Exponential and Logarithmic Expressions

OTHER NOTES

An _____ is an expression of the form _____
where b is a _____ or a number raised to an _____ x. In the
expression, b is called the _____ and x indicates the _____ the
base is used as a _____ in repeated multiplication.

EXAMPLES 1–3 Evaluate each exponential expression for the given
value of x.
5^x; $x = 2$

4^x; $x = 3$

2^x; $x = 5$

Any non-zero number raised to the zero power is equal to _____.

Any number raised to the power of 1 is equal to _____.

EXAMPLE 4 Evaluate the exponential expression for $x = 0$ and $x = 1$.
3^x

A negative exponent leads to a _____. That is, $x^{-n} =$.

EXAMPLES 5 & 6 Evaluate each exponential expression for the given value
of x.
2^x; $x = -1$

6^x; $x = -2$

What is a logarithm?

How is a logarithm, base b of a positive number x represented?

475

EXAMPLES 7–10 Evaluate each logarithmic expression.

$\log_2 4$

$\log_5 5$

$\log_3 \dfrac{1}{27}$

$\log_4 1$

CONCEPT CHECK:

How would the expression 2^x be evaluated for $x = 6$?

GUIDED EXAMPLES:

Evaluate each exponential expression for the given value of x.

1. 3^x; $x = 4$

2. 4^x; $x = 1$

3. 4^x; $x = -3$

4. Evaluate the logarithmic expression. $\log_3 81$

Name: _____   Date: _____

Instructor: _____   Section: _____

Notebook 38.4
Graphing Exponential Functions

An _____ is an expression of the form _____ where b _____ and x is a real number.
The number b is called the _____ of the function.

EXAMPLE 1 Graph the exponential function.

$f(x) = 2^x$

x	$f(x)$
-2	
-1	
0	
1	
2	
3	

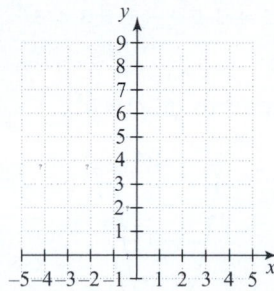

What is an asymptote?

What is the asymptote for every exponential function of the form $f(x) = b^x$?

What is the domain for an exponential function $f(x) = b^x$?

What is the range for an exponential function $f(x) = b^x$?

Note: What happens to the graph when the base is greater than 1?

EXAMPLE 2 Graph the exponential function.

$f(x) = \left(\dfrac{1}{2}\right)^x$

x	$f(x)$
-3	
-2	
-1	
0	
1	
2	

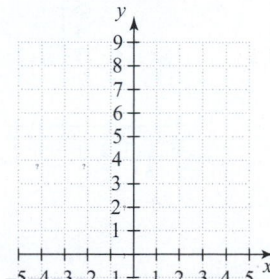

Define the number e.

What is the exponential function with base e?

477

EXAMPLE 3 Graph the exponential function.

$$f(x) = e^x$$

x	$f(x)$
-2	
-1	
0	
1	
2	

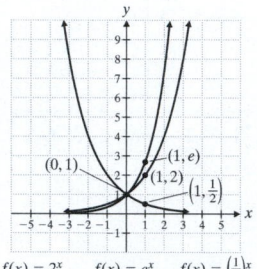

The graph of an exponential function of the form $f(x) = b^x$,

1. Passes through the points

   _____ and _____.

2. If $b > 1$, the graph is …

   If $b < 1$, the graph is …

$f(x) = 2^x$ $f(x) = e^x$ $f(x) = \left(\frac{1}{2}\right)^x$

CONCEPT CHECK:

Is the graph of $f(x) = 0.9^x$ increasing or decreasing?

GUIDED EXAMPLES: Graph the exponential functions.

1. $f(x) = 3^x$

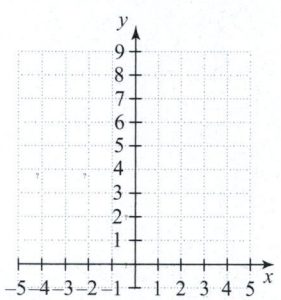

2. $f(x) = \left(\frac{1}{3}\right)^x$

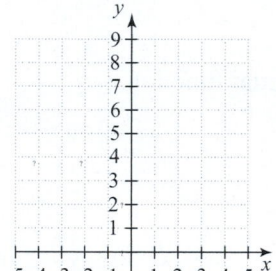

3. $f(x) = e^{-x}$

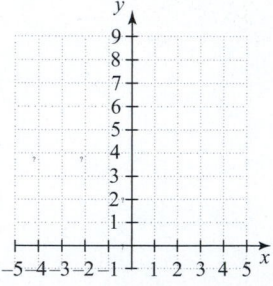

478

Notebook 38.5
Converting Between Exponential and Logarithmic Forms

The _____, base b, of a positive number x, is the _____ (exponent) to which the base b must be raised to produce _____, and is represented by _____.

What is the property that relates exponential and logarithmic forms?

EXAMPLE 1 Write in logarithmic form. $81 = 3^4$
Identify x, b and y in $x = b^y$.

Rewrite in log form.

EXAMPLES 2–4 Write in logarithmic form.
$49 = 7^2$

$512 = 8^3$

$\dfrac{1}{100} = 10^{-2}$

EXAMPLE 5 Write in exponential form. $2 = \log_5 25$
Identify x, b and y in $y = \log_b x$.

Rewrite in exponential form.

EXAMPLES 6–8 Write in exponential form.

$$\frac{1}{2} = \log_{16} 4$$

$$0 = \log_{17} 1$$

$$-4 = \log_{10}\left(\frac{1}{10,000}\right)$$

CONCEPT CHECK:

What exponential equation is the same as the logarithmic equation $y = \log_b x$?

GUIDED EXAMPLES:

Write in logarithmic form.

1. $36 = 6^2$

2. $\frac{1}{32} = 2^{-5}$

Write in exponential form.

3. $3\log_5 125$

4. $5 = \log_3 243$

Notebook 38.6
Graphing Logarithmic Functions

An _____ is an expression of the form
_____ where b _____ and x is a
_____ real number.

The logarithmic equation _____ is the same as the exponential
equation _____, where $b > 0$ and $b \neq 1$.

Write the steps for graphing logarithmic functions.

1.

2.

Note. What is the first step to convert a logarithmic function to exponential
form?

EXAMPLE 1 Graph the logarithmic function.

$f(x) = \log_2 x$

x	y
	-2
	-1
	0
	1
	2

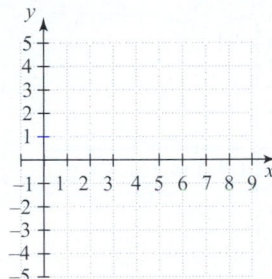

What is the asymptote for every logarithmic function of the form
$f(x) = \log_b x$?

What is the domain for a logarithmic function?

What is the range for a logarithmic function?

EXAMPLE 2 Graph the exponential function.

$f(x) = \log_{1/3} x$

x	y
	-1
	0
	1
	2

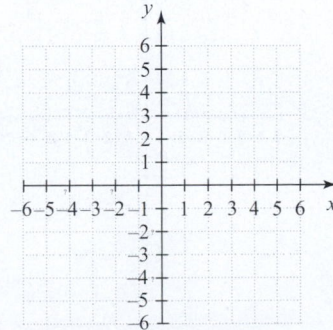

481

What is the inverse function of the logarithmic function?

What is the inverse function of the exponential function?

What does that say about the graphs of each?

EXAMPLE 3 Graph each function on the same set of axes.

$f(x) = \log_2 x$ and $f(x) = 2^x$

$y = 2^x$

x	y
-1	
0	
1	
2	

$f(x) = \log_2 x$

x	y
	-1
	0
	1
	2

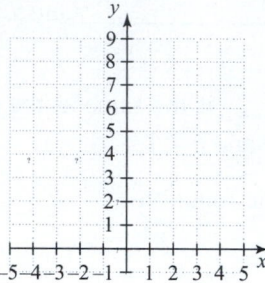

CONCEPT CHECK:

What is the asymptote of a logarithmic function of the form $f(x) = \log_b x$?

GUIDED EXAMPLES:

1. Graph the logarithmic function.

    $f(x) = \log_3 x$

    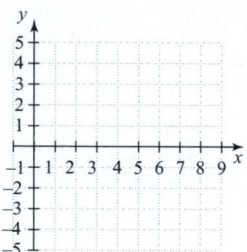

2. Graph the logarithmic function.

    $f(x) = \log_{1/2} x$

    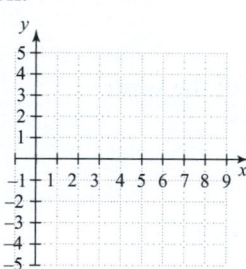

3. Graph each function on the same set of axes.

    $f(x) = \log_6 x$ and $f(x) = 6^x$

    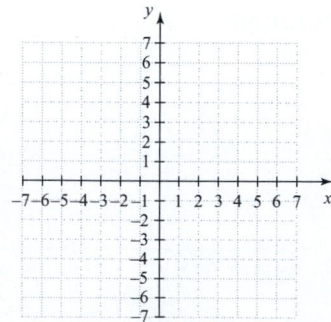

482

Notebook 39.1
Properties of Logarithms

What does the Logarithm Property of a Product state?

That is, the log of the _____ is equal to the _____ of the logs.

EXAMPLES 1 & 2 Write each logarithm as a sum of logarithms.

$\log_3 AB$

$\log_5 (7 \cdot 11)$

EXAMPLE 3 Write the expression as a single logarithm.

$\log_3 14 + \log_3 x$

Caution! $(\log_b M)(\log_b N) \neq$

What does the Logarithm Property of a Quotient state?

The log of the _____ is equal to the _____ of the logs.

EXAMPLE 4 Write the logarithm as a difference of logarithms.

$\log_3 \left(\dfrac{29}{7} \right)$

EXAMPLE 5 Write the expression as a single logarithm.

$\log_3 45 - \log_3 9$

Caution! $\dfrac{\log_b M}{\log_b N} \neq$

What does the Logarithm Property of a Number raised to a Power say?

EXAMPLES 6 & 7 Express each logarithm as a product.

$\log_5 b^{10}$

$\log_8 \sqrt{w}$

EXAMPLE 8 Write the expression as a single logarithm.

$\dfrac{1}{3}\log_b x + 2\log_b w - 3\log_b z$

EXAMPLE 9 Write the logarithm as a sum or difference of logarithms.

$\log_b \left(\dfrac{x^4 y^3}{z^2} \right)$

Caution! The base of the logarithm(s) must be the _____.

CONCEPT CHECK:

The logarithmic expression $\log_b MN$ can be rewritten as a _____ of logarithms.

GUIDED EXAMPLES:

1. Write the logarithm as a sum of logarithms. $\log_8 XZ$

2. Write the expression as a single logarithm. $\log_b 4 + \log_b x$

3. Express the logarithm as a product. $\log_b A^{-2}$

4. Write the expression as a single logarithm. $2\log_b 7 + 3\log_b y - \dfrac{1}{2}\log_b z$

484

Notebook 39.2
Common and Natural Logarithms

The _____, base *b*, of a positive number *x*, is the power
(_____) to which the _____ *b* must be raised to produce *x*, and
is represented by the expression _____.

What are the most frequently used bases of logarithms?

What is a common logarithm?

EXAMPLES 1–3 On a scientific or graphing calculator, approximate the
following values.
$\log 7.32$

$\log 73.2$

$\log 0.314$

The most useful base for logarithms and exponential expressions is
_____.

Define natural logarithm.

EXAMPLES 4–6 On a scientific or graphing calculator, approximate the
following values.
$\ln 7.21$

$\ln 72.1$

$\ln 0.0356$

Note: log 1 = _____ and ln 1 = _____. Why?

CONCEPT CHECK:
What is the base of a common logarithm?

GUIDED EXAMPLES: On a scientific or graphing calculator, approximate the following values.

1. log 4.36

2. log 78,500

3. ln 0.0793

4. ln 136,000

Notebook 39.3
Change of Base of Logarithms

What does it mean to evaluate $\log_b x$?

Examples:
$\log_2 8 = \qquad\qquad \log_7 49 = \qquad\qquad \log_3 11 =$

Why is the change of base formula necessary?

What is the change of base formula for logarithms?

EXAMPLE 1 Evaluate the logarithm using common logarithms. Round
to three decimal places.
$\log_3 11$
Identify the base b and x.
Use the change of base formula.
Use a calculator to approximate the log.

EXAMPLES 2 & 3 Evaluate each logarithm using common logarithms.
Round
to three decimal places.
$\log_{15} 12$

$\log_7 5.12$

487

EXAMPLE 4 Evaluate the logarithm using natural logarithms. Round to three decimal places.

$\log_4 0.0057739$

Identify the base b and x.

Use the change of base formula.

Use a calculator to approximate the log.

EXAMPLES 5 & 6 Evaluate each logarithm using natural logarithms. Round to three decimal places.

$\log_{21} 436$

$\log_6 0.315$

How can you check your answers?

CONCEPT CHECK:

Why is the change of base formula used?

GUIDED EXAMPLES: Evaluate the logarithm using common logarithms. Round to three decimal places.

1. $\log_{17} 18$

2. $\log_5 50.3$

Evaluate the logarithm using natural logarithms. Round to three decimal places.

3. $\log_{30} 913$

4. $\log_7 0.004462$

Notebook 39.4
Solving Simple Exponential Equations and Applications

OTHER NOTES

State the Property of Exponential Equations.

Write the steps for solving exponential equations.

1. Get the _____ on both sides of the equation, if possible.

2. Set the _____ on each side _____ to each other.

3. If necessary, _____ the equation from Step 2.

EXAMPLES 1 & 2 Solve the exponential equations.
$2^x = 4$

$3^x = 1$

EXAMPLES 3 & 4 Solve the exponential equations.
$$2x = \frac{1}{16}$$

$8^{3x-1} = 64$

State and explain the compound interest formula.

EXAMPLE 5 If a young married couple invests $5000 in a mutual fund that pays 16% interest compounded annually, how much will they have in 3 years?
Identify the known values.
Substitute into the compound interest formula.
Simplify.

State and explain the variable compound interest formula.

EXAMPLE 6 If we invest $8000 in a fund that pays 15% annual interest compounded monthly, how much will we have in the account after 6 years?
Identify the known values.
Substitute the values into the formula.
Simplify.

State and explain the exponential decay formula.

EXAMPLE 7 The decay constant for americium 241 is $k = -0.0016008$.
If 10 milligrams (mg) of americium 241 is sealed in a laboratory container today, how much will still be present in 50 years?

CONCEPT CHECK:
In the exponential decay formula $A = Ce^{kt}$ for a radioactive element, what quantity does C represent??

GUIDED EXAMPLES: Solve the exponential equation.

1. $2^x = 8$

2. $3^x = \dfrac{1}{27}$

3. $4^{-x} = \dfrac{1}{16}$

4. $5^{x+1} = 125$

Name: _____ Date: _____

Instructor: _____ Section: _____

Notebook 39.5
Solving Exponential Equations and Applications

State the property for taking the logarithm of both sides.

Write the steps for solving exponential equations.

1.

2.

3.

Recall the properties for logarithms:

$\log_b MN =$

$\log_b \left(\dfrac{M}{N} \right) =$

$\log_b M^p =$

EXAMPLE 1 Solve the exponential equation. Leave your answer in exact form.

$2^x = 7$

Take the log of each side.

Solve the equation.

Check the solution.

EXAMPLE 2 Solve the exponential equation. Approximate your answer to the nearest thousandth.

$3^x = 7^{x-1}$

Take the log of each side.

Solve the equation.

Check the solution.

Note: What do you do if the exponential equation involves e raised to a power?

Note: For any base b, $\log_b b =$ _____, so $\log 10 =$ _____ and $\ln e =$ _____.

491

EXAMPLE 3 Solve the exponential equation. Approximate your answer to the nearest ten-thousandth.

$e^{2.5x} = 8.42$

Take the natural log of each side.
Solve the equation.

Check the solution.

If a principal amount P is invested at an interest rate r, compounded _____, the amount of _____ A, accumulated after t years is $A =$ _____.

EXAMPLE 4 If P dollars is invested in an account that earns interest at 12% compounded annually, the amount available after t years is $A = P(1 + 0.12)t$. How many years will it take for $300 in this account to grow to $1500? Round your answer to the nearest whole year?

State and explain the Growth Equation.

EXAMPLE 5 If the world population is eight billion people and it continues to grow at the current rate of 2% per year, how many years will it take for the population to increase to fourteen billion people?

CONCEPT CHECK: How can $\log 3^x = \log 15$ be rewritten?

GUIDED EXAMPLES:
1. Solve the exponential equation. Leave your answer in exact form.

    $3^x = 5$

2. Solve the exponential equation. Approximate your answer to the nearest ten-thousandth.

    $20.98 = e^{3.6x}$

3. If P dollars is invested in an account that earns interest at 8% compounded annually, the amount available after t years is $A = P(1 + 0.08)^t$. How many years will it take for $4000 in this account to grow to $10,000? Round your answer tot he nearest whole year.

4. The wildlife management team in one region of Alaska has determined that the black bear population is growing at the rate of 4.3% per year. This region currently has approximately 1300 black bears. A food shortage will develop if the population reaches 5000. If the growth rate remains unchanged, how many years will it take for this food shortage problem to occur?

Name: _____   Date: _____

Instructor: _____   Section: _____

Notebook 39.6
Solving Simple Logarithmic Equations

OTHER NOTES

The logarithmic equation $y = \log_b x$ is the same as _____.
Write the steps for solving logarithmic equations.
1.

2.

EXAMPLE 1 Solve the logarithmic equation.
$\log_2 x = 4$

EXAMPLES 2–4 Solve the logarithmic equations.
$3 = \log_3 x$

$\log_{10} x = -3$

$\log_{25} x = \dfrac{1}{2}$

What is the first step in solving an exponential equation?

EXAMPLES 5 & 6 Solve the logarithmic equations.
$\log_7 343 = y$

$\log_8 \dfrac{1}{64} = y$

493

How do you solve an exponential equation for the base?

EXAMPLES 7 & 8 Solve the logarithmic equations.

$2 = \log_b 121$

$\log_b 81 = 4$

CONCEPT CHECK:

To solve for x in $4 = \log_3 x$, what exponential equation needs to be rewritten?

GUIDED EXAMPLES: Solve the logarithmic equations.

1. $5 = \log_{10} x$

2. $\log_4 64 = y$

3. $2 = \log_b 144$

4. $\log_b 125 = 3$

Notebook 39.7
Solving Logarithmic Equations and Applications

Write the steps for solving logarithmic equations.
1. If there is more than one logarithmic term, …

2. Get _____ logarithmic term on …

3. _____ the equation to an _____ equation using …

4. _____ the equation.

5. _____ the solution(s).

Recall the properties for logarithms:
$\log_b MN =$

$\log_b\left(\dfrac{M}{N}\right) =$

$\log_b M^p =$

EXAMPLE 1 Solve for x. $\log 5 = 2 - \log(x + 3)$
Isolate the log.

Convert to an exponential equation.

Solve the equation.

Check the solution(s).

EXAMPLE 2 Solve for x. $\log_3(x + 6) - \log_3(x - 2) = 2$
Isolate the log.

Convert to an exponential equation.

Solve the equation.

Check the solution(s).

State the Logarithmic Property of Equality.

Write the steps for solving logarithmic equations with only logarithmic terms.

1.

2.

3.

4.

Caution! It is not possible to take the logarithm of _____.

EXAMPLE 3 Solve for x. $\log(x+6) + \log(x+2) = \log(x+20)$

Write each side with a single log.

Use Logarithmic Property of Equality.

Solve the equation.

Check the solution.

EXAMPLE 4 The magnitude of an earthquake (amount of energy released) is measured by the formula $R = \log\left(\dfrac{I}{I_0}\right)$, where I is the intensity of the earthquake and I_0 is the minimum measurable intensity. The 1964 earthquake in Anchorage, Alaska, had a magnitude of 8.4. The 1906 earthquake in Taiwan had a magnitude of 7.1. How many times more intense was the Anchorage earthquake than the Taiwan earthquake?

CONCEPT CHECK: What is the logarithmic property of equality?

GUIDED EXAMPLES:

Solve for x.

1. $\log(x+5) = 2 - \log 5$

2. $\log_{11}(x+7) = \log_{11}(x-3) + 1$

3. $\log_5(2x) - \log_5(x-3) = 3\log_5 2$

4. The magnitude of an earthquake (amount of energy released) is measured by the formula, $R = \log\left(\dfrac{I}{I_0}\right)$, where I is the intensity of the earthquake and I_0 is the minimum measurable intensity. the 1989 earthquake in San Francisco had a magnitude of 7.1. The 1993 earthquake in Northridge, California had a magnitude of 6.8. How many times more intense was the San Francisco earthquake than the Northridge earthquake?

Name: _____   Date: _____

Instructor: _____   Section: _____

Notebook A.1
Using a Graphing Calculator

Why use a graphing calculator?

What does the viewing window refer to?

What is the "standard" window?   Give the settings.

What is the "square" window?   How do you correct this?   Give the settings.

EXAMPLE 1 Graph $y = -\dfrac{1}{2}x + 3$ on a graphing calculator. Write the keystrokes used.

Most calculators only understand equations that are …

EXAMPLE 2 Graph $3x + 2y = 6$ on a graphing calculator. Write the steps below.

EXAMPLE 3 Graph $f(x) = x^2 - 5$ on a graphing calculator and find the y-intercept. Write the steps.

Another limitation is the inability to understand …

EXAMPLE 4 Solve for y and graph. $x = y^2 - 3$. Write the steps below.

EXAMPLE 5 Graph $y = x^2 + 3x - 10$ on a graphing calculator and find the vertex. Write the steps.

EXAMPLE 6 Graph $y = -x^2 - 2x + 5$ on a graphing calculator and find domain and range.
Steps.

What is the difference between dot mode and connected mode on a calculator?

Caution! When graphing rational functions, where is this problem most often missed?

EXAMPLE 7 Graph the function in connected mode and identify the domain and the y-intercept.

$$f(x) = \frac{9x - 12}{(x+3)(x-4)}$$

Steps.

Caution! Are the vertical lines part of the graph?

OTHER NOTES

497

EXAMPLE 8 Graph the function in dot mode and identify the domain and the y-intercept.

$$f(x) = \frac{6x}{x^2 - 25}$$

Steps.

What are the ways you can locate the values of x for which a function is not defined?

How do you determine if two rational expressions are equivalent?

EXAMPLES 9 & 10 In each of the following, use your graphing calculator to determine if y_1 is equivalent to y_2.

$$y_1 = \frac{2x+1}{x-3} - \frac{5x}{5x+15} \quad \text{and} \quad y_2 = \frac{x+1}{x+3}$$

$$y_1 = \frac{3x}{2x-3} + \frac{3x+6}{2x^2+x-6} \quad \text{and} \quad y_2 = \frac{3x+3}{2x+3}$$

EXAMPLE 11 Graph the function and identify the domain and range.

$$f(x) = \sqrt{x+5}$$

EXAMPLE 12 Solve $\frac{1}{2}x^2 + 2x - 4 = \sqrt{x+2}$ by finding the point of

intersection of $y_1 = \frac{1}{2}x^2 + 2x - 4$ and $y_2 = \sqrt{x+2}.$ Write the steps in the space below.

What other features can be used to find the intersection?

EXAMPLE 13 Solve $\sqrt{4-x} = x-2$ by finding the point of intersection of $y_1 = \sqrt{4-x}$ and $y_2 = x-2.$ Write the steps in the space below.

How can you input an exponent other than 2?

EXAMPLE 14 Solve $\frac{1}{4}x^3 + 1 = \frac{1}{2}x + \frac{25}{4}$ by finding the point of intersection of

$y_1 = \frac{1}{4}x^3 + 1$ and $y_2 = \frac{1}{2}x + \frac{25}{4}.$ Write the steps in the space below.

CONCEPT CHECK: Which of the following equations can be entered into the graphing portion of a calculator EXACTLY as shown?
$x = y^2 + 3$ or $y = x^2 + 10x + 1$

498

Notebook A.2
Algebra of Functions

How are functions represented?

What does $f(x)$ mean?
What is the rule for algebraic operations on functions for:
Sum of two functions?

Difference of two functions?

Product of two functions?

Quotient of two functions?

EXAMPLES 1 & 2 For $f(x) = 2x$ and $g(x) = x + 1$, perform the indicated operations.
$(f + g)(x)$

$(f - g)(x)$

EXAMPLES 3 & 4 For $f(x) = 4x + 3$ and $g(x) = x + 12$, perform the indicated operations.
$(f \cdot g)(-3)$

$\left(\dfrac{f}{g}\right)(-12)$

EXAMPLES 5 & 6 For $p(x) = 3x + 9$ and $q(x) = x + 122x - 11$, perform the indicated operations.
$(p + q)(2)$

$(pq)(x)$

EXAMPLES 7 & 8 For $f(x) = 7x^2 + 30x + 8$ and $g(x) = x + 4$, perform the indicated operations.

$(f - g)(-3)$

$\left(\dfrac{f}{g}\right)(x)$

The _____ of a function is the set of _____ that can be substituted into a function, and the _____ is the set of possible _____ for the function.

How do you get the domain for combined functions? (Which one is different?)

$(f + g)(x)$

$(f - g)(x)$

$(f \cdot g)(x)$

$\left(\dfrac{f}{g}\right)(x)$

CONCEPT CHECK: For $f(x) = -2x$ and $g(x) = 3x - 4$, how will $(f - g)(x)$ be performed?

GUIDED EXAMPLES:

1. For $f(x) = 2x + 8$ and $g(x) = -6x$, evaluate $(g - f)(x)$.

2. For $f(x) = 2x^2 + 8x - 3$ and $g(x) = 9x + 2$, evaluate $(f + g)(2)$.

3. For $b(x) = 2x^2 + 8x - 3$ and $d(x) = 3x^2$, evaluate $(db)(-1)$.

4. For $f(x) = 2x^2 + 8x - 3$ and $h(x) = 3x^2$, evaluate $\left(\dfrac{f}{h}\right)(3)$.

500

Name: _____   Date: _____

Instructor: _____   Section: _____

Notebook A.3
Transformations of Functions

What are some types of transformations of functions?
Compare the graphs of $f(x) = x$ and $g(x) = x + 3$.
What do you notice from the table of values?
What do you notice about the graphs?

What do we mean by a "vertical shift"?
What are the properties of vertical shifts?
1.
2.
EXAMPLE 1 Graph $h(x)$ by applying a vertical shift to the graph of $f(x)$.
$f(x) = x^2$ and $h(x) = x^2 + 2$

Graph $f(x)$.

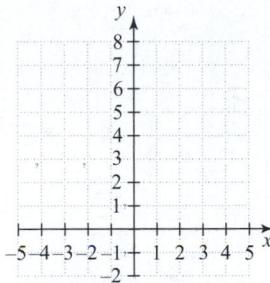

Determine the necessary shift to graph $h(x)$.
Graph $h(x)$ by applying the vertical shift.

Compare the graphs of $h(x) = x^2$ and $r(x) = (x + 1)^2$.
What do you notice about the graphs?

What do we mean by a "horizontal shift"?
What are the properties of horizontal shifts?
1.
2.
EXAMPLE 2 Graph $p(x)$ by applying a horizontal shift to the graph of $f(x)$.
$f(x) = |x|$ and $p(x) = |x - 3|$
Graph $f(x)$.

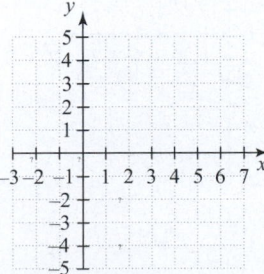

Determine the necessary shift to graph $p(x)$.
Graph $h(x)$ by applying the horizontal shift.

501

Is it possible to have both a vertical and horizontal shift?

EXAMPLE 3 Graph $h(x)$ by applying the appropriate shifts to the graph of $f(x)$.

$f(x) = x^3$ and $h(x) = (x - 3)^3 - 2$

Graph $f(x)$.

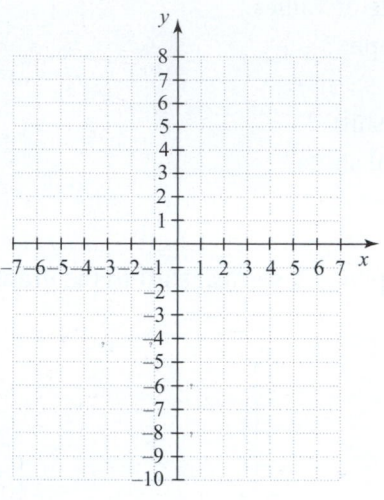

Determine the necessary shift to graph $h(x)$.

Graph $h(x)$ by applying the necessary shifts.

What is meant by a reflection of a graph?

EXAMPLE 4 Graph $g(x)$ by transforming the graph of $f(x)$.

$f(x) = x^2$ and $g(x) = -x^2$

Graph $f(x)$.

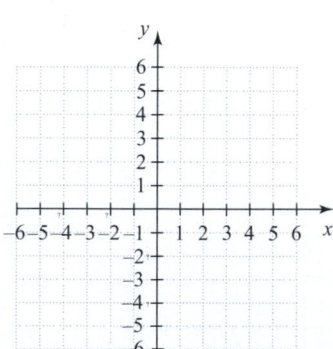

Reflect the graph to graph $g(x)$.

Summarize the transformations of functions.

502

Name: _____   Date: _____

Instructor: _____   Section: _____

CONCEPT CHECK: Describe the graph of $f(x) + k$.

GUIDED EXAMPLES:

1. Graph $k(x)$ by applying the vertical shift to the graph of $g(x)$.

   $g(x) = 2x$ and $k(x) = 2x - 3$

   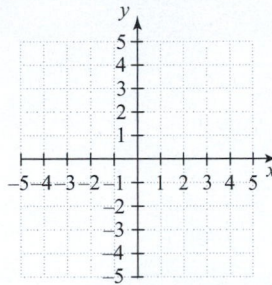

2. Graph $P(x)$ by applying a horizontal shift to the graph of $f(x)$.

   $f(x) = x^2$ and $P(x) = (x + 2)^2$

   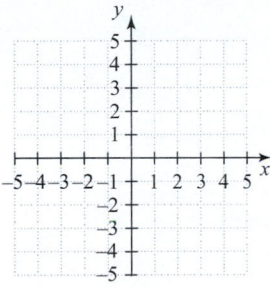

3. Graph $g(x)$ by applying the appropriate shift to the graph of $f(x)$.

   $f(x) = x^3$ and $h(x) = (x + 4)^3 + 3$

   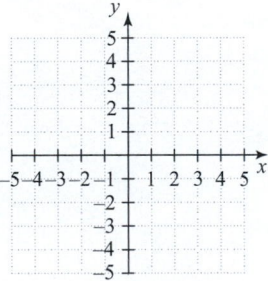

4. Graph $g(x)$ by transforming the graph of $f(x)$. $f(x) = |x|$ and $g(x) = -|x|$

   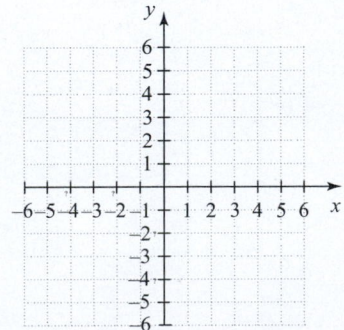

503

Notebook A.4
Piecewise Functions

Why would you need a piecewise function?

What is a piecewise function?

Write the steps for evaluating a piecewise function.
1.

2.

Caution! Can an x-value have more than one y-value?

EXAMPLE 1 If $f(x) = \begin{cases} 2x & \text{if} \quad x < 2 \\ x^2 & \text{if} \quad x \geq 2 \end{cases}$ find the following.

$f(3)$
Determine which domain the value fits into.

Substitute into the corresponding equation.

EXAMPLE 2 If $f(x) = \begin{cases} 2x+3 & \text{if} \quad x \leq 0 \\ -x-1 & \text{if} \quad x > 0 \end{cases}$ find the following.

$f(2)$

$f(-6)$

$f(0)$

EXAMPLE 3 Susan's weekly salary (in dollars) is given by the piecewise
function $f(x) = \begin{cases} 10x & \text{if} \quad 0 \le x \le 40 \\ 20x - 400 & \text{if} \quad x > 40 \end{cases}$ where x is the number of hours
worked per week. Determine how much money Susan will make if she works
the following number of hours in a week.

30 hours

60 hours

The _____ of a number is the distance between than number
and zero on a number line.

EXAMPLE 4 Write the following absolute value function as a piecewise
function.

$f(x) = |x|$

$f(10)$

$f(-5)$

Sketch the graph of this function.

CONCEPT CHECK: What is the first step in evaluating a piecewise
function?

GUIDED EXAMPLES:

1. If $f(x) = \begin{cases} 2x - 5 & \text{if} \quad x < 1 \\ x + 1 & \text{if} \quad x \ge 1 \end{cases}$, find $f(-3)$.

2. If $f(x) = \begin{cases} -2x + 7 & \text{if} \quad x \le 0 \\ x - 8 & \text{if} \quad x > 0 \end{cases}$, find the following. $f(-4)$

Erin's monthly cell phone bill is a flat rate of $20 plus 10¢ per minute used. The
amount of her cell phone bill is given by the piecewise function

$f(x) = \begin{cases} 20 & \text{if} \quad 0 \le x < 400 \\ 20 + 0.10x & \text{if} \quad x \ge 400 \end{cases}$, where x is the number of minutes she

uses. Determine how much Erin's cell phone bill will be if she uses the
following number of minutes in a month.

3. 200 minutes

4. 400 minutes

Name: _____   Date: _____

Instructor: _____   Section: _____

Notebook A.5
Graphing Piecewise Functions

OTHER NOTES

A _____ is a function whose definition changes depending on the value of the _____.

What are the two steps for evaluating a piecewise function?
1.

2.

What are the steps for graphing a piecewise function?

What is the graph of a function of the form $f(x) = b$?

EXAMPLE 1 Graph the piecewise function.

$$f(x) = \begin{cases} 4 & \text{if} \quad x < -3 \\ -2 & \text{if} \quad x \geq -3 \end{cases}$$

Make a table of values for each piece.

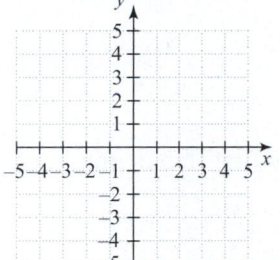

Graph each piece.

EXAMPLE 2 Graph the piecewise function.

$$f(x) = \begin{cases} 2x+3 & \text{if} \quad x \leq 0 \\ -x-1 & \text{if} \quad x > 0 \end{cases}$$

Make a table of values for each piece.

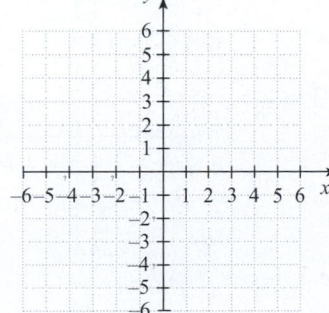

Graph each piece.

507

EXAMPLE 3 Graph the piecewise function.

$$f(x) = \begin{cases} 2x & \text{if} \quad x \le 2 \\ x^2 & \text{if} \quad x > 2 \end{cases}$$

Make a table of values for each piece.

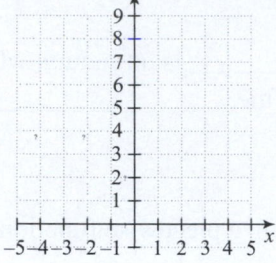

Graph each piece.

EXAMPLE 4 Graph the piecewise function.

$$f(x) = \begin{cases} -x & \text{if} \quad x \le -1 \\ 2x & \text{if} \quad -1 < x \le 2 \\ 4 & \text{if} \quad x \ge 2 \end{cases}$$

Make a table of values for each piece.

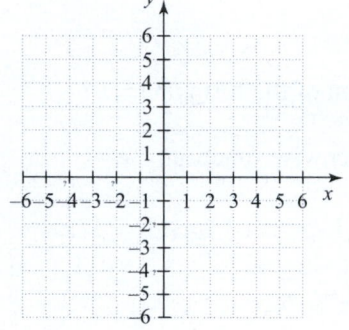

Graph each piece.

CONCEPT CHECK: What is the first step in graphing

$$f(x) = \begin{cases} x+2 & \text{if} \quad x < 2 \\ 2x & \text{if} \quad x \ge 2 \end{cases}?$$

GUIDED EXAMPLES: Graph the piecewise function.

1. $f(x) = \begin{cases} x+2 & \text{if} \quad x < 1 \\ 2x-1 & \text{if} \quad x \ge 1 \end{cases}$

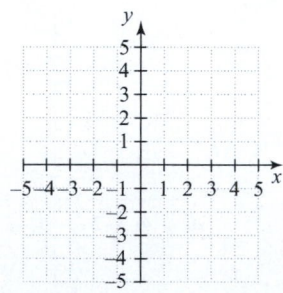

508

2. $f(x) = \begin{cases} -1 & \text{if} \quad x \le 0 \\ -3 & \text{if} \quad x \ge 2 \end{cases}$

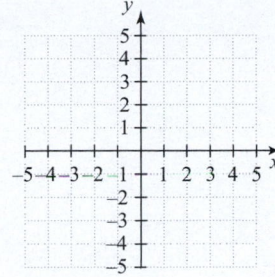

3. $f(x) = \begin{cases} 4x-4 & \text{if} \quad x < 2 \\ -x+6 & \text{if} \quad x \ge 2 \end{cases}$

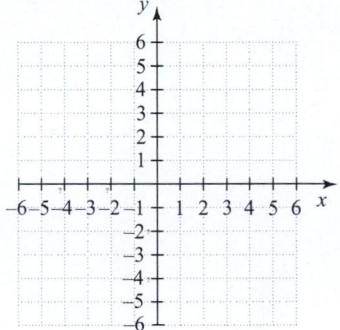

4. $f(x) \begin{cases} -2x-2 & \text{if} \quad x \le -2 \\ \dfrac{1}{2}x+3 & \text{if} \quad -2 < x \le 2 \\ -x+6 & \text{if} \quad x > 2 \end{cases}$

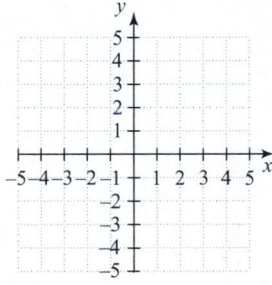

Name: _____   Date: _____

Instructor: _____   Section: _____

Notebook B.1
Systems of Linear Inequalities

A _____ is two or more linear inequalities graphed on the same set of axes.

The _____ to a system of linear inequalities is the _____ of the solution sets of the individual inequalities.

EXAMPLE 1 Graph the solution to the system of linear inequalities.

$y \le 4$

$x > 2$

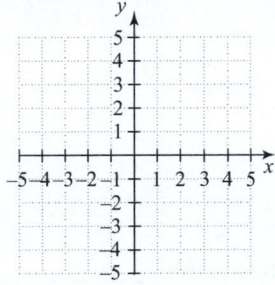

Graph the first inequality.

Graph the second inequality.

Graph both graphs on the same graph.

EXAMPLE 2 Graph the solution to the system of linear inequalities.

$y < -\dfrac{4}{3}x + 3$

$y \ge \dfrac{1}{2}x - 2$

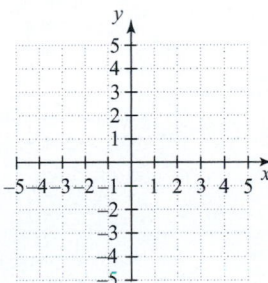

EXAMPLE 3 Graph the solution to the system of linear inequalities.

$y \ge 2x + 3$

$4x - 2y > 4$

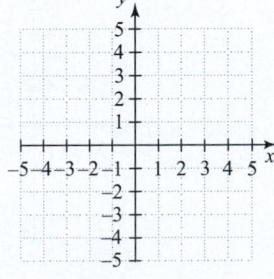

EXAMPLE 4 Graph the solution to the system of linear inequalities. Find and label all the points of intersection.

$$x + y \le 3$$
$$x - y \le 1$$
$$x \ge -1$$

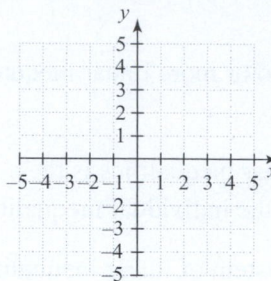

Graph the first inequality.
Graph the second inequality.
Graph the third inequality.
Mark the solution.
Label the intersection points.

CONCEPT CHECK:

The solution to a system of linear inequalities is the _____ of the solution sets of the individual inequalities.

GUIDED EXAMPLES: Graph the solutions to these systems of inequalities.

1. $x \le 0$
    $y > -1$

2. $y < -\dfrac{2}{3}x + 4$
    $y > x - 1$

3. $5y \ge 5x + 10$
    $x - y \ge 0$

4. $y < x$
    $x + y \ge 1$
    $x \le 4$

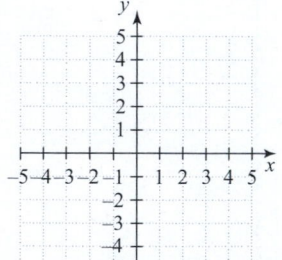
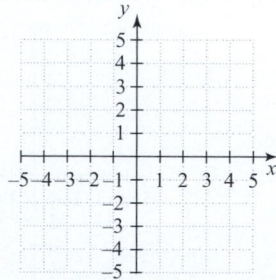

512

Name: _____   Date: _____

Instructor: _____   Section: _____

Notebook B.2
Systems of Non-Linear Equations

A _____ is two or more equations with the same variables.

The _____ to a system of equations in two variables is an _____ which is a solution to every equation in the system.

What does this mean related to a graph?

What are some different ways we can solve a system of equations?

Explain the graphing method.

What is a system of non-linear equations?

How many solutions will there be?

EXAMPLE 1 Solve by graphing.

$y = x^2$

$y = x + 6$

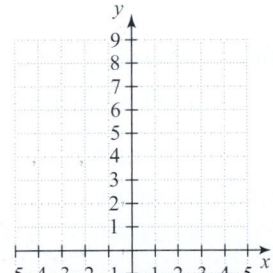

Explain the substitution method.

When will you want to solve a system using substitution?

EXAMPLE 2 Solve using the substitution method.

$y = x^2$

$y = 2x + 15$

513

Explain the elimination method.

When is the elimination method usually the easiest method?

EXAMPLE 3 Solve using the elimination method.
$$2x^2 - 5y^2 = -2$$
$$3x^2 + 2y^2 = 35$$

EXAMPLE 4 Solve by any method.
$$x^2 + y^2 = 4$$
$$y = x^2 - 2$$

EXAMPLE 5 Solve by any method.
$$x^2 + y^2 = 4$$
$$y = 5 - x$$

CONCEPT CHECK:
Can the following system of equations be solved by the elimination method?
Why or why not?
$$x^2 - 2y^2 = 1$$
$$2x^2 + y = 10$$

GUIDED EXAMPLES:

1. Solve the system of equations by graphing. $f(x) = -2x^2 + 7$
$$f(x) = -2x + 3$$

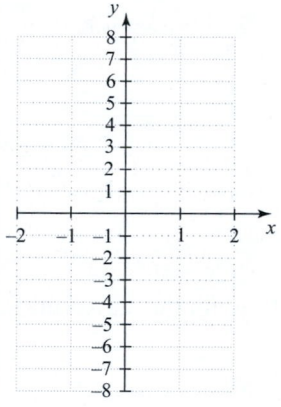

2. Solve the system of equations by substitution. $y = x^2$
$$y = x + 12$$

3. Solve the system of equations by elimination. $x^2 + 4y^2 = 13$
$$x^2 - 3y^2 = -8$$

4. Solve the system of equations by any method. $x^2 + y^2 = 0$
$$x - y = 6$$

514

Notebook B.3
Matrices and Determinants

What is a matrix?

What symbol do we use to indicate a matrix?

Give some examples of matrices and give their sizes.

How do you give the dimension of a matrix?

How can matrices be used to represent systems of equation?

EXAMPLE 1 Write the coefficients of the variables in the system of equations as a matrix.

$$5y = 10$$
$$2x - 6y = 17$$

What is a square matrix?

What is a determinant?

How do we indicate a determinant?

What is the rule for calculating the value of a 2×2 determinant $\begin{vmatrix} a & c \\ b & d \end{vmatrix}$?

EXAMPLES 2 & 3 Find the determinant of each matrix.

$$\begin{bmatrix} -6 & 2 \\ -1 & 4 \end{bmatrix}$$

$$\begin{bmatrix} 0 & -3 \\ -2 & 6 \end{bmatrix}$$

What is the rule for evaluating a 3×3 determinant?

EXAMPLE 4 Evaluate the determinant. $\begin{vmatrix} 4 & 1 & 2 \\ 3 & -1 & 0 \\ 1 & 2 & 3 \end{vmatrix}$

What do we mean by the "minor" of an element?

515

EXAMPLES 5 & 6 Find the minor of 6 and the minor of –3 in the determinant.

$$\begin{vmatrix} 6 & 1 & 2 \\ -3 & 4 & 5 \\ -2 & 7 & 8 \end{vmatrix} \quad \text{minor of 6}$$
$$\begin{vmatrix} 6 & 1 & 2 \\ -3 & 4 & 5 \\ -2 & 7 & 8 \end{vmatrix} \quad \text{minor of } -3$$

Write the procedure for evaluating a 3×3 determinant using expansion by minors.

EXAMPLE 7 Evaluate the determinant by expanding it by minors of elements in the first column.

$$\begin{vmatrix} 2 & 3 & 6 \\ 4 & -2 & 0 \\ 1 & -5 & -3 \end{vmatrix}$$

What is an array of signs for a 3×3 determinant?

What is it used for?

EXAMPLE 8 Evaluate the determinant by expanding it by minors of elements in the second row.

$$\begin{vmatrix} 1 & -3 & 5 \\ 2 & -2 & 0 \\ -6 & 4 & -7 \end{vmatrix}$$

CONCEPT CHECK: What is the size of the matrix? $\begin{bmatrix} 1 & 9 & -8 \\ 7 & 4 & 4 \end{bmatrix}$?

GUIDED EXAMPLES:

1. Write the coefficients of the variables in the system of equations as a matrix.

    $-5x + y = -6$
    $x - y = 0$

2. Find the determinant of the matrix. $\begin{bmatrix} -7 & 3 \\ -4 & -2 \end{bmatrix}$

3. Find the determinant. $\begin{vmatrix} 3 & -4 & -1 \\ -2 & 1 & 3 \\ 0 & 1 & 4 \end{vmatrix}$

4. Evaluate the determinant by expanding it by minors of elements in the second column.

    $$\begin{vmatrix} 2 & 0 & -2 \\ -1 & 0 & 2 \\ 3 & 4 & 3 \end{vmatrix}$$

516

Name: _____ Date: _____

Instructor: _____ Section: _____

Notebook B.4
Solving Systems of Linear Equations Using Matrices

What is an augmented matrix?

What is the purpose of the vertical bar in an augmented matrix?

What are the rules for matrix row operations?
1.

2.

3.

What is the desired form(s) of an augmented matrix?

EXAMPLE 1 Solve the system of equations using a matrix.
$$4x - 3y = -13$$
$$x + 2y = 5$$

Write the augmented matrix.

Use row operations to achieve the augmented matrix.

Write as a system of equations.

Solve this system using substitution.

517

EXAMPLE 2 Solve the system of equations using a matrix.

$2x + 3y - z = 11$

$x + 2y + z = 12$

$3x - y + 2z = 5$

Write the augmented matrix.

Use row operations to achieve the augmented matrix.

Write as a system of equations.

Solve this system using substitution.

CONCEPT CHECK: Write the augmented matrix for the following system of equations.

$x - 5y = 3$

$3x - 4y = 7$

GUIDED EXAMPLES:

1. Solve the system of equations using a matrix.

$x + 5y = -9$

$4x - 3y = -13$

2. Solve the system of equations using a matrix.

$x - 2y - 3z = 4$

$2x + 3y + z = 1$

$-3x + y - 2z = 5$

518

Name: _____   Date: _____

Instructor: _____   Section: _____

Notebook B.5
Cramer's Rule

What is the value of the 2 × 2 determinant $\begin{vmatrix} a & c \\ b & d \end{vmatrix}$?

What is the name of the method that uses determinants to solve a system of linear equations?

What does Cramer's Rule state?

The solution to   is $x =$ and $y =$,

where D 0 and $D =$, $D_x =$ and $D_y =$.

EXAMPLE 1 Solve the system by Cramer's Rule.

$-3x + y = 7$
$-4x - 3y = 5$

Find D.

Find D_x.

Find D_y.

Solve for x and y. $x =$ $y =$

Write the solution as an ordered pair.

EXAMPLE 2 Solve the system by Cramer's Rule.

$x + 3y = 6$
$2x + y = 7$

Find D.

Find D_x.

Find D_y.

Solve for x and y. $x =$ $y =$

Write the solution as an ordered pair.

What does Cramer's Rule look like for a system of three equations with three variables?

EXAMPLE 3 Solve the system by Cramer's Rule.

$2x - y + z = 6$
$3x + 2y - z = 5$
$2x + 3y - 2z = 1$

Find D.

Find D_x.

Find D_y.

Find D_z.

Solve for x, y, and z. $x =$ $y =$ $z =$

Write the solution as an ordered triple.

Note: What happens if $D = 0$ in Cramer's Rule?

CONCEPT CHECK: The determinants for a system of two linear equations in two variables has been found to be $D = 10$, $D_x = 20$, and $D_y = -30$. What is the solution to the system of equations?

GUIDED EXAMPLES:

1. Solve the system by Cramer's Rule. $x + 2y = 8$
$2x + y = 7$

2. Solve the system by Cramer's Rule. $5x + 4y = 10$
$-x + 2y = 12$

520

Notebook C.1
Sets

What is a set?

What is the symbol for "is an element of the set"?

What are natural numbers?
What letter represents the set of natural numbers?
How do we represent sets?
The elements in the set are contained in _____ and separated by
_____.

What does it mean that the set of natural numbers is infinite?

EXAMPLES 1 & 2 List all the elements in each set.
Set X is the set of all natural numbers between 4 and 10.

Set Y is the set of all natural numbers between 2 and 7, inclusive.

What is a finite set?

What is set-builder notation?
When is this useful?
Example.

EXAMPLES 3 & 4 Write in set-builder notation.
$A = \{a, e, i, o, u\}$

Set B is the set of natural numbers between –6 and 0.

Caution!! What are the symbols used for an empty set?
Why can $\{\emptyset\}$ not be used?

Name two operations that can be used with sets.
1.   2.

What is set intersection?

What is set union?

If $A = \{a, b, c, d, e\}$ and $B = \{a, e, i, o, u\}$,
what is $A \cap B$?
What is $A \cup B$?

EXAMPLE 5 Find the intersection and the union of the sets C and D.
$C = \{$odd numbers between 1 and 11, inclusive$\}$
$D = \{$multiplies of 3 between 1 and 15, inclusive$\}$

Find $C \cap D$.

Find $C \cup D$.

What is a subset?

$A = \{$consonants$\}$ and $B = \{$letters of the alphabet$\}$
How is a subset written?

EXAMPLES 6 & 7 Determine if each statement is true or false. Give the
reason.
If $A = \{$odd numbers$\}$ and $B = \{$integers$\}$, then $A \subseteq B$.

If $A = \{$multiples of 3$\}$ and $B = \{$multiples of 4$\}$, then $A \subseteq B$.

CONCEPT CHECK:

Describe the union of $A = \{1, 2, 3, 5, 7, 9, 10, 11\}$ and
$B = \{2, 3, 4, 6, 8, 10, 11\}$.

GUIDED EXAMPLES:

1. List the elements the set.

   Set X is the set of all natural numbers between -3 and 6.

2. Find the intersection and the union of the sets.

   $A = \{-1, 2, 3, 5, 8\}$ and $B = \{-2, -1, 3, 5|$

For the following problems, $A = \{1, 2, 3, 4, 5...\}$, $B = \{10, 20, 30, 40...\}$, and
$C = \{1, 2, 3, 4, 5\}$

3. True or false: $B \subseteq A$

4. True or false: $B \cap C = \{10, 20, 30, 40\}$

Notebook C.2
The Midpoint Formula

What is the midpoint of a line segment?

What is the formula for finding the midpoint of a line segment
whose endpoints are (x_1, y_1) and (x_2, y_2)?

Caution!! What is the difference between the distance between two points and
the midpoint?

EXAMPLE 1 Find the midpoint of the line segment whose endpoints are
the following points. (1, 4) and (3, 2)

EXAMPLES 2–4 Find the midpoint of each line segment whose endpoints
are the following points.
(–5, 0) and (9, –4)

$\left(\dfrac{1}{2}, -\dfrac{3}{4}\right)$ and $\left(-\dfrac{1}{2}, \dfrac{1}{4}\right)$

(20, –6.5) and (–13, –7.3)

523

CONCEPT CHECK: What is the midpoint of a line segment?

GUIDED EXAMPLES:

1. Find the midpoint of the line segment whose endpoints are the following points: $(3, 6)$ and $(7, 2)$

2. Find the midpoint of the line segment whose endpoints are the following points: $(8, -3)$ and $(0, -9)$

3. Find the midpoint of the line segment whose endpoints are the following points: $\left(-\dfrac{1}{3}, \dfrac{2}{5}\right)$ and $\left(-\dfrac{2}{3}, \dfrac{1}{5}\right)$

4. Find the midpoint of the line segment whose endpoints are the following points: $(12.5, -2.3)$ and $(-3.5, 2.3)$

Name: _____   Date: _____

Instructor: _____   Section: _____

Notebook C.3
Surface Area

OTHER NOTES

Let's review.
Write the formula for finding the area of each of the following.

Rectangle

Square

Parallelogram

Triangle

Equilateral Triangle

Circle

What are the 4 basic types of 3-dimensional figures?

What is surface area?
What unit is used to measure surface area?

EXAMPLES 1 & 2 Find the number of surfaces and describe their shapes.

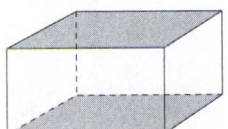 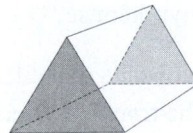

EXAMPLES 3 & 4 Find the number of surfaces and describe their shapes.

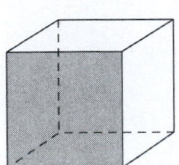 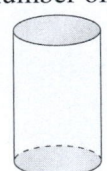

EXAMPLE 5 Find the surface area.

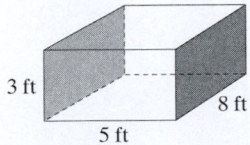

3 ft   8 ft
5 ft

EXAMPLE 6 Find the surface area of this cube.

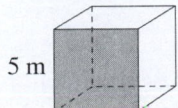

5 m

525

EXAMPLE 7 A sheet of aluminum is rolled to make a cylindrical aluminum can with a height of 10 inches and a radius of 2 inches. How much aluminum is needed to make this can?

What is the formula for finding the surface area of a sphere with radius r?

EXAMPLE 8 The planet Mars has a diameter of approximately 6800 km. Find the surface area of Mars. (Assume the planet is a sphere.)

What is the formula for the area of an equilateral triangle with sides b?

EXAMPLE 9 A cardboard tube used for mailing posters is in the shape of a triangular prism, with each end shaped like an equilateral triangle. The sides of the triangle are each 5 inches, and the tube is 24 inches long. Find the surface area of the tube. Round your answer to the nearest hundredth.

In a right triangle, if a and b represent the legs and c represents the hypotenuse, what is the relationship between the legs and the hypotenuse?

EXAMPLE 10 While building a model ship, Saul wants to fold a sheet of brass into the shape of a right triangular prism. The triangular ends of the prism are right triangles, 9 mm high by 40 mm wide. The prism will have a length of 200 mm. How much brass is needed to make the prism?

CONCEPT CHECK: Which figure has three surfaces?

GUIDED EXAMPLES: Find the surface area of each figure**.**

1. Find the surface area.   2. Find the surface area.   3. Find the surface area.

Use 3.14 for π.

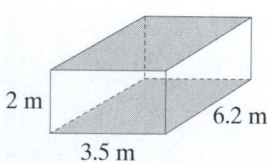

2 m
6.2 m
3.5 m

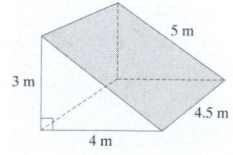

5 m
3 m
4.5 m
4 m

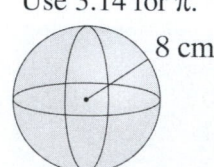

8 cm

Name: _____ Date: _____

Instructor: _____ Section: _____

Notebook C.4
Synthetic Division

Let's review polynomial division.

Note: What is meant by missing terms?

In synthetic division, we can ignore _____ and concentrate on the
_____ of each term.

Write the steps of dividing using synthetic division.
1. Write the polynomial in _____, and fill in any
   _____ of x.

2. _____ the division using the _____
   of the terms only, and the _____ of the constant term of the
   divisor.

3. Write the _____ of the first term in the far left of the
   _____.

4. Multiply the _____ by the number from the _____
   in the bottom row and _____ the product to the _____ number in the
   _____.

5. Write the _____ in the _____ row.

6. Repeat this until _____.

7. The last _____ in the result area is the _____.

8. Write the answer as a _____, with the numbers in the
   result area as the _____ of the terms, in _____
   _____. The degree of the quotient is _____ than the degree
   of the _____. Write the _____ as a fraction over the
   polynomial _____.

EXAMPLE 1 Divide using synthetic division. $(3x^3 + 7x^2 - 4x + 3) \div (x + 3)$

527

EXAMPLE 2 Divide using synthetic division.
$(3x^4 - 21x^3 + 31x^2 - 25) \div (x - 5)$

CONCEPT CHECK:

What is the FIRST step to divide polynomials using synthetic division?

GUIDED EXAMPLES:

1. Divide using synthetic division. $(4x^3 - 11x^2 - 20x + 5) \div (x - 4)$

2. Divide using synthetic division. $(3x^4 + 8x - 10) \div (x + 2)$

Notebook C.5
Balancing a Checking Account

What is a debit?

What is a credit?

What is the procedure for finding the amount of money in a checking account?

EXAMPLE 1 As of 7/11, Alex Murgaton had a balance of $1657.38 in his checking account. On 7/12, Alex wrote check #177 to Shop Market for $122.96. On 7/13, he used his debit card at Chelsea's Cinema for $16 and at Bob's Convenience Store for $10.97. On 7/14, his weekly check of $590.27 from work was directly deposited into his account, and he withdrew $60 from an ATM. On 7/15, his automatic payments of $30.02 to the Electric Company, $17.76 to the Gas Company, and $650.00 to Sky High Apartments were paid. Record the credits and debits in Alex's check register, and then find his ending balance.

CHECK REGISTER *Alex Murgaton*						20 *11*	
CHECK NO.	DATE	DESCRIPTION OF TRANSACTION	PAYMENT/ DEBIT (−)	✓	DEPOSIT/ CREDIT (+)	BALANCE $ 1657	38
177	7/12	*Shop Market*	122 96	✓		1534	42

What does it mean to "balance a checkbook"?

What are the steps for balancing a checkbook?
1.

2.

3.

EXAMPLE 2 Morgan O'Malley calculated her checkbook balance, but it does not match her bank's records. Her bank states that as of 4/8, she should have $600.53 in her checking account. Look at Morgan's check register and identify and fix any errors that she might have made. If she made no errors, then indicate that the bank must have made an error.

CHECK REGISTER *Morgan O'Malley* 20 **12**

CHECK NO.	DATE	DESCRIPTION OF TRANSACTION	PAYMENT/ DEBIT (−)	✓	DEPOSIT/ CREDIT (+)	BALANCE $568 17
303	4/2	Pet Market	44 02	✓		524 15
	4/3	Direct Deposit		✓	988 15	1512 30
	4/5	River Place Apartments		✓	767 99	2280 80
304	4/5	Violin Lessons	30 00	✓		2250 29
	4/8	Al's Food Market (debit card)	113 78	✓		2126 51

EXAMPLE 3 Balance the monthly statement for Lorence Seafood using the restaurant's check register and bank statement.

CHECK REGISTER *Lorence Seafood* 20 **12**

CHECK NO.	DATE	DESCRIPTION OF TRANSACTION	PAYMENT/ DEBIT (−)	✓	DEPOSIT/ CREDIT (+)	BALANCE $2607 33
268	2/2	Shoreline Fishery	613 07	✓		
269	2/10	Tia Sullivan	430 00			
270	2/10	Moe Cormorant	430 00	✓		
271	2/10	Evan Enriquéz	620 00			
	2/14	Deposit		✓	1770 00	
272	2/20	Raymund Property Management	575 00	✓		
273	2/24	Wegstaff Appliance Repair	447 74	✓		
	3/1	Deposit		✓	1300 00	

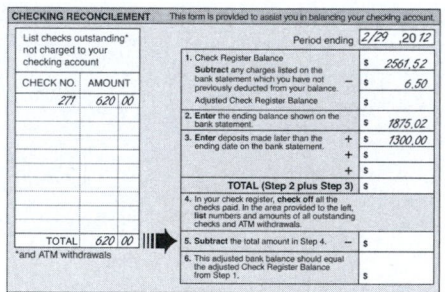

Bank Statement: LORENCE SEAFOOD   2/1/2012 to 2/29/2012

Beginning Balance $2607.33
Ending Balance $1875.02

Checks cleared by the bank

#268 $613.07   #270* $430.00   #273 $447.74
#269 $430.00   #272 $575.00
*Indicates that the next check in the sequence is outstanding (hasn't cleared).

Deposits
2/14 $1770.00

Other withdrawals
Service charge $6.50

CHECKING RECONCILEMENT This form is provided to assist you in balancing your checking account.

Period ending 2/29 , 20 12

List checks outstanding* not charged to your checking account

CHECK NO.	AMOUNT
271	620 00
TOTAL	620 00

*and ATM withdrawals

1. Check Register Balance — $ 2561.52
   Subtract any charges listed on the bank statement which you have not previously deducted from your balance. − $ 6.50
   Adjusted Check Register Balance $
2. **Enter** the ending balance shown on the bank statement. $ 1875.02
3. **Enter** deposits made later than the ending date on the bank statement. + $ 1300.00
   + $
   + $
   + $
   TOTAL (Step 2 plus Step 3) $
4. In your check register, **check off** all the checks paid. In the area provided to the left, **list** numbers and amounts of all outstanding checks and ATM withdrawals.
5. **Subtract** the total amount in Step 4. − $
6. This adjusted bank balance should equal the adjusted Check Register Balance from Step 1. $

CONCEPT CHECK:

Name some reasons why a checkbook register would be inaccurate.

Name: _____   Date: _____

Instructor: _____   Section: _____

Notebook C.6
Determining the Best Deal when Purchasing a Vehicle

What is the purchase price of a vehicle?

What is a down payment?

What is the amount financed?

Note: How do you find a $p\%$ of a number?
How do you write a percent as a decimal?

EXAMPLE 1 Scott purchased a truck that was on sale for $29,999 in a city that has a 6.5% sales tax and a 2% license fee.
a. Find the amount of sales tax and license fee Scott paid.

b. Scott also bought an extended warranty for $1650. Find the purchase price of the truck.

EXAMPLE 2 The purchase price of a car Lisa wants to buy is $15,999. In order to qualify for the loan on the car, Lisa must make a down payment of 20% of the purchase price.
a. Find the amount of the down payment.

b. Find the amount financed.

What is total cost?

531

EXAMPLE 3 James went to two dealerships to find the best deal on a sports car he plans to purchase. From which dealer should James buy the car so that the total cost of the car is the least expensive?

Dealership 1	Dealership 2
Purchase price: $36,999	Purchase price: $36,999
Financing option: 3% financing with $5000 down payment	Financing option: 5% financing with no down payment
Monthly payments: $686.65 per month for 48 months	Monthly payments: $647.48 per month for 60 months

Find the total cost of the car at Dealership 1.

Find the total cost of the car at Dealership 2.

Which is the best deal?

CONCEPT CHECK:
What describes the purchase price of a vehicle?

GUIDED EXAMPLES:
1. Levi bought a car that was on sale for $11,999 in a town that has a 4% sales tax and a 3% license fee.
   a. Find the amount of sales tax and license fee Levi paid.

   b. Levi also bought an extended warranty for $1325. Find the purchase price of the truck.

2. The purchase price of a car Sarah plans to buy is $8599. In order to qualify for the loan on the car, Lisa must make a down payment of 16% of the purchase price. **Note:** Sarah lives in a state without sales tax.

   a. Find the amount of the down payment.

   b. Find the amount financed.

3. Keira went to two dealerships to find the best deal on a sports car she plans to purchase. From which dealer should Keira buy the car so that the total cost of the car is the least expensive.

Dealership 1	Dealership 2
Purchase price: $34,999	Purchase price: $33,999
Financing option: 3% financing with $5000 down payment	Financing option: 5% financing with no down payment
Monthly payments: $643.73 per month for 48 months	Monthly payments: $594.98 per month for 60 months